Air Pollution Handbook

Air Pollution Handbook

Kylan Wilkins

SYRAWOOD
PUBLISHING HOUSE

New York

Published by Syrawood Publishing House,
750 Third Avenue, 9th Floor,
New York, NY 10017, USA
www.syrawoodpublishinghouse.com

Air Pollution Handbook
Kylan Wilkins

International Standard Book Number: 978-1-64740-022-4 (Hardback)

Cataloging-in-Publication Data

Air pollution handbook / Kylan Wilkins.
    p. cm.
Includes bibliographical references and index.
ISBN 978-1-64740-022-4
1. Air--Pollution. 2. Air quality. 3. Pollution. 4. Atmospheric deposition. I. Wilkins, Kylan.
TD883 .A47 2020
363.739 2--dc23

TABLE OF CONTENTS

PREFACE

The contamination of Earth's atmosphere through harmful or extreme quantities of particles, gases or biological molecules is known as air pollution. It can have various adverse effects on the health of humans such as causing diseases, allergies or in extreme cases, even death. Other living organisms such as animals and food crops can also be harmed by it. Air pollution can be caused due to both human and natural processes. A few examples of the different pollutants are carbon monoxide gas, ash from volcanoes and sulfur dioxide. Some of the alternatives which can be adopted in order to reduce air pollution are using biofuel for airplanes, using electric motor vehicles and using renewable sources of energy to generate electricity. This book covers in detail some existent theories and innovative concepts revolving around air pollution. It presents this complex problem in the most comprehensible and easy to understand language. With its detailed analyses and data, this book will prove immensely beneficial to professionals and students involved in this area at various levels.

Given below is the chapter wise description of the book:

Chapter 1- The introduction of harmful or excessive quantities of substances into the atmosphere of the Earth is known as air pollution. A few of such substances are carbon dioxide, carbon monoxide, CFC and radon. All the diverse aspects of air pollution have been briefly introduced in this chapter.

Chapter 2- Air pollution can be caused due to anthropogenic reasons such as waste incineration and smoking as well as natural reasons like volcanic eruptions, radon emissions and methane emissions. The chapter closely examines these key causes of air pollution to provide an extensive understanding of the subject.

Chapter 3- There are various materials which are classified as air pollutants such as nitrogen oxides, ozone, sulfur dioxide, carbon monoxide, volatile organic compounds, mercury, ammonia, lead and chlorofluorocarbons. The topics elaborated in this chapter will help in gaining a better perspective about these air pollutants as well as short-lived climate pollutants.

Chapter 4- Air pollution has a diverse range of ill effects which affect human health, domestic animals, environment and ozone layer. It can also lead to global warming, acid rain and wind erosion. The chapter closely examines these effects of air pollution to provide an extensive understanding of the subject.

Chapter 5- Air pollution modeling refers to the usage of mathematical theories in order to understand and predict the behavior or pollutants in the air. Measurement of air pollution includes air quality monitoring and maintaining the air quality index. The topics elaborated in this chapter will help in gaining a better perspective about the diverse aspects of air pollution modeling and measurement.

Chapter 6- Air pollution control refers to the techniques which are used to remove the harmful substances which pollute the air. The chapter closely examines several focus areas related to air pollution control such as the equipment used for air pollution control and the diverse ways to stop air pollution.

Indeed, my job was extremely crucial and challenging as I had to ensure that every chapter is informative and structured in a student-friendly manner. I am thankful for the support provided by my family and colleagues during the completion of this book.

Kylan Wilkins

Chapter 1

Introduction to Air Pollution

The introduction of harmful or excessive quantities of substances into the atmosphere of the Earth is known as air pollution. A few of such substances are carbon dioxide, carbon monoxide, CFC and radon. All the diverse aspects of air pollution have been briefly introduced in this chapter.

Steel is an alloy of iron, carbon and other elements. It is an important material in engineering as it is cheap and has a high tensile strength. The amount of carbon in the alloy and the form in which it is present regulates steel's characteristics. The chapter on steel and steel making offers an insightful focus, keeping in mind the complex subject matter.

Air pollution is a form of pollution that refers to the contamination of the air, irrespective of indoors or outside. A physical, biological or chemical alteration to the air in the atmosphere can be termed as pollution. It occurs when any harmful gases, dust, smoke enters into the atmosphere and makes it difficult for plants, animals, and humans to survive as the air becomes dirty.

Air pollution can further be classified into two sections- visible air pollution and invisible air pollution. Another way of looking at air pollution could be any substance that holds the potential to hinder the atmosphere or the well being of the living beings surviving in it. The sustainment of all things living is due to a combination of gases that collectively form the atmosphere; the imbalance caused by the increase or decrease in the percentage of these gases can be harmful to survival.

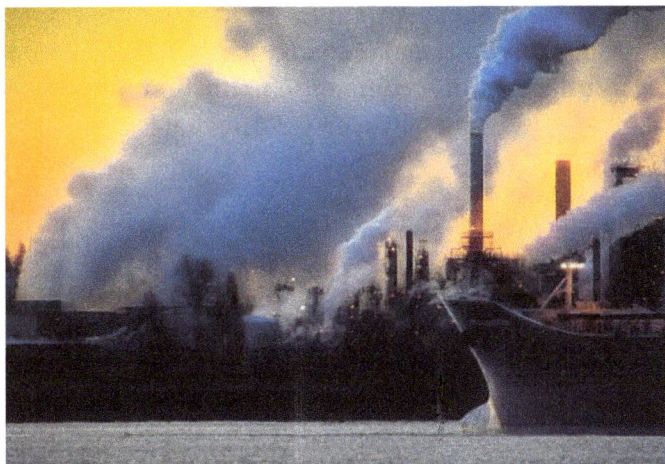

The Ozone layer considered crucial for the existence of the ecosystems on the planet is depleting due to increased pollution. Global warming, a direct result of the increased imbalance of gases in the atmosphere has come to be known as the biggest threat and challenge that the contemporary world has to overcome in a bid for survival.

Types of Pollutants

In order to understand the causes of Air pollution, several divisions can be made.

- Primarily air pollutants can be caused by primary sources or secondary sources. The pollutants that are a direct result of the process can be called primary pollutants. A classic example of a primary pollutant would be the sulfur-dioxide emitted from factories.

- Secondary pollutants are the ones that are caused by the intermingling and reactions of primary pollutants. Smog created by the interactions of several primary pollutants is known to be as a secondary pollutant.

Various Causes of Air pollution

The Burning of Fossil Fuels

Sulfur dioxide emitted from the combustion of fossil fuels like coal, petroleum and other factory combustibles are one the major cause of air pollution. Pollution emitting from vehicles including trucks, jeeps, cars, trains, airplanes cause an immense amount of pollution. We rely on them to fulfill our daily basic needs of transportation.

But, their overuse is killing our environment as dangerous gases are polluting the environment. Carbon Monoxide caused by improper or incomplete combustion and generally emitted from vehicles is another major pollutant along with Nitrogen Oxides, that is produced from both natural and man-made processes.

Agricultural Activities

Ammonia is a very common by product from agriculture-related activities and is one of the most hazardous gases in the atmosphere. Use of insecticides, pesticides, and fertilizers in agricultural activities has grown quite a lot. They emit harmful chemicals into the air and can also cause water pollution.

Exhaust from Factories and Industries

Manufacturing industries release a large amount of carbon monoxide, hydrocarbons, organic compounds, and chemicals into the air thereby depleting the quality of air. Manufacturing industries can be found at every corner of the earth and there is no area that has not been affected by it. Petroleum refineries also release hydrocarbons and various other chemicals that pollute the air and also cause land pollution.

Mining Operations

Mining is a process wherein minerals below the earth are extracted using large equipment. During the process dust and chemicals are released in the air causing massive air pollution. This is one of the reasons which is responsible for the deteriorating health conditions of workers and nearby residents.

Household cleaning products, painting supplies emit toxic chemicals in the air and cause air pollution. Have you ever noticed that once you paint the walls of your house, it creates some sort of smell which makes it literally impossible for you to breathe?

Suspended particulate matter popular by its acronym SPM, is another cause of pollution. Referring to the particles afloat in the air, SPM is usually caused by dust, combustion, etc.

Disastrous Effects of Air pollution

Respiratory and Heart Problems

The effects of air pollution are alarming. They are known to create several respiratory and heart conditions along with Cancer, among other threats to the body. Several million are known to have died due to direct or indirect effects of Air pollution. Children in areas exposed to air pollutants are said to commonly suffer from pneumonia and asthma.

Global Warming

Another direct effect is the immediate alterations that the world is witnessing due to global warming. With increased temperatures worldwide, increase in sea levels and melting of ice from colder regions and icebergs, displacement and loss of habitat have already signaled an impending disaster if actions for preservation and normalization aren't undertaken soon.

Acid Rain

Harmful gases like nitrogen oxides and sulfur oxides are released into the atmosphere during the burning of fossil fuels. When it rains, the water droplets combine with these air pollutants becomes acidic and then falls on the ground in the form of acid rain. Acid rain can cause great damage to human, animals, and crops.

Eutrophication

Eutrophication is a condition where a high amount of nitrogen present in some pollutants gets developed on sea's surface and turns itself into algae and adversely affect fish, plants and animal species. The green colored algae that are present on lakes and ponds is due to the presence of this chemical only.

Effect on Wildlife

Just like humans, animals also face some devastating effects of air pollution. Toxic chemicals present in the air can force wildlife species to move to a new place and change their habitat. The toxic pollutants deposit over the surface of the water and can also affect sea animals.

Depletion of the Ozone Layer

Ozone exists in the Earth's stratosphere and is responsible for protecting humans from harmful ultraviolet (UV) rays. Earth's ozone layer is depleting due to the presence of chlorofluorocarbons, hydro chlorofluorocarbons in the atmosphere. As the ozone layer will go thin, it will emit harmful rays back on earth and can cause skin and eye related problems. UV rays also have the capability to affect crops.

When you try to study the sources of Air pollution, you enlist a series of activities and interactions that create these pollutants. There are two types of sources that we will take a look at Natural sources and Man-made sources.

Natural sources of pollution include dust carried by the wind from locations with very little or no green cover, gases released from the body processes of living beings (Carbon dioxide from humans during respiration, Methane from cattle during digestion, Oxygen from plants during Photosynthesis). Smoke from the combustion of various inflammable objects, volcanic eruptions, etc. along with the emission of polluted gases also makes it to the list of natural sources of pollution.

While looking at the man-made contributions towards air pollution, smoke again features as a prominent component. The smoke emitted from various forms of combustion like in biomass, factories, vehicles, furnaces, etc. Waste used to create landfills generate methane that is harmful in several ways. The reactions of certain gases and chemicals also form harmful fumes that can be dangerous to the well being of living creatures.

Solutions for Air Pollution

Use Public Mode of Transportation

Encourage people to use more and more public modes of transportation to reduce pollution. Also, try to make use of carpooling. If you and your colleagues come from the same locality and have same timings you can explore this option to save energy and money.

Conserve Energy

Switch off fans and lights when you are going out. A large number of fossil fuels are burnt to produce electricity. You can save the environment from degradation by reducing the number of fossil fuels to be burned.

Understand the Concept of Reduce, Reuse and Recycle

Do not throw away items that are of no use to you. In-fact reuse them for some other purpose. For e.g. you can use old jars to store cereals or pulses.

Emphasis on Clean Energy Resources

Clean energy technologies like solar, wind and geothermal are on high these days. Governments of various countries have been providing grants to consumers who are interested in installing solar panels for their home. This will go a long way to curb air pollution.

Use energy Efficient Devices

CFL lights consume less electricity as against their counterparts. They live longer, consume less electricity, lower electricity bills and also help you to reduce pollution by consuming less energy.

Several attempts are being made worldwide on personal, industrial and governmental levels to curb the intensity at which air pollution is rising and regain a balance as far as the proportions of the foundation gases are concerned. This is a direct attempt at slacking Global warming. We are seeing a series of innovations and experiments aimed at alternate and unconventional options to reduce pollutants. Air pollution is one of the larger mirrors of man's follies, and a challenge we need to overcome to see a tomorrow.

Indoor Air Pollution

Indoor air pollution occurs when certain air pollutants from particles and gases contaminate the air of indoor areas. These air pollutants can cause respiratory diseases or even cancer. Removing the air pollutants can improve the quality of your indoor air.

Millions of people around the world prepare their meals using traditional methods (i.e. wood, charcoal, coal, dung, crop wastes) on open fires. Such inefficient practices can increase the amount of air pollutants inside the home and can also cause serious health problems. According to WHO, 4.3 million people a year die from the exposure to household air pollution.

This type of pollution is significantly more dangerous due to how concentrated the air is in indoor environments. According to recent findings, over 2 million deaths occur every single year due to indoor air pollution. So what can we do about it? That is the question that many ask themselves every single day. Before you can fully comprehend the effects of indoor air pollution you must first be able to understand the causes of it as well as what we can do to improve our quality of air both indoors and outdoors.

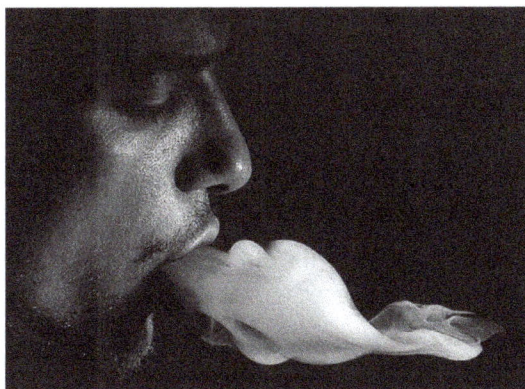

Causes of Indoor Air Pollution

Toxic products, inadequate ventilation, high temperature and humidity are few of the primary causes of indoor air pollution in our homes.

- Asbestos is the leading cause of indoor air pollution. Asbestos can be found in various materials used commonly in the automotive industry as well as home construction. They are most commonly found in coatings, paints, building materials, and ceiling and floor tiles. You won't find asbestos as often as you used because newer products do not contain asbestos. However, if you have an old home that was constructed a long time ago, the risks for asbestos are much greater than that of a newer home. Asbestos has been banned in the US and is no longer being used.

- Formaldehyde is another leading cause of indoor air pollution. It is no longer produced in the United States due to its ban in 1970 but can still be found in paints, sealants, and wood floors.

- Radon which can be found underneath your home in various types of bedrock and other building materials can also be a cause of indoor air pollution. Radon can get into the walls of your home and put both you and your family at risk.

- Tobacco smoke that comes from outdoor and indoor areas can also be an indoor air pollutant.

- Many contaminants that grow in damp environments can be brought in from outdoor areas. These contaminants such as mildew, mold, bacteria, dust mites, as well as animal dander, can come into the home and make you sick.

- There are many objects that you have in your home that also cause indoor air pollution. Objects such as wood stoves, space heaters, and fireplaces, all put out carbon monoxide as well as nitrogen dioxide. There are still billions of people who use these types of fuels to heat their homes on a daily basis.

- Other household products such as varnishes, paints, and certain cleaning products can also emit pollution into the air that you breathe inside your home.

Serious Effects of Indoor Air Pollution

Effects of indoor air pollution can be life threatening. Kids and old age people are more prone to the after-effects of indoor air pollution.

- If Asbestos is found in your home it can cause you very serious health problems such as lung cancer, asbestosis, mesothelioma, and various other types of cancers.

- If contaminants such as animal dander, dust mites or other bacteria get into the home there will also be some serious effects from them. You will start to experience asthma symptoms, throat irritation, flu, and other types of infectious diseases.

- If lead is found in the home it can also be severely life threatening. It can cause brain and nerve damage, kidney failure, anemia, and a defective cardiovascular system.

- Formaldehyde, one of the most common indoor air pollutants, can also cause health problems. You may experience irritation of the throat, eyes, and nose, as well as allergic reactions. There have been a number of cases where it has also caused cancer.

- Tobacco smoke causes individuals to experience severe respiratory irritation, pneumonia, bronchitis, emphysema, heart disease, as well as lung cancer.

- Chemicals such as those that are used in certain cleaning agents and paints can cause you to experience a loss of coordination, liver, brain, and kidney damage, as well as a number of types of cancer.

- If you use gas stoves in your home it can cause respiratory infections and damage and irritation to the lungs.

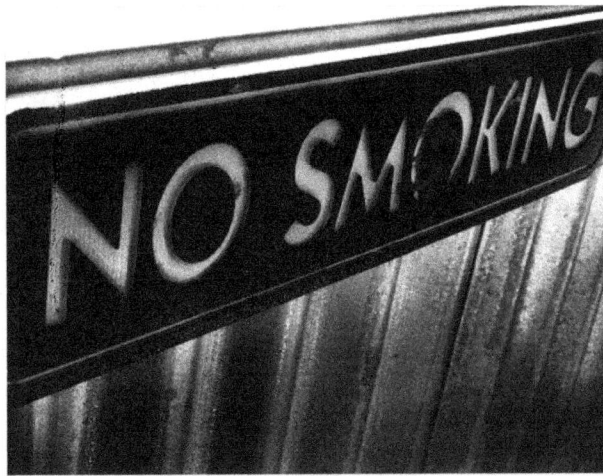

Ways to Improve Indoor Air Quality

- Smoking is one of the most common types of indoor air pollution. The best thing to do is to quit smoking and make your home anti-smoking zone. The less smoke that is emitted into the air the less chance of one of the listed effects happening to someone that you love. Smoking is a leading cause of cancer. Lung cancer is the most common form of cancer caused by smoking.

- Make sure you check the ingredients on any of your cleaning supplies to make sure they are environmentally friendly. Do your homework on what is considered to be a dangerous ingredient. You can also find an environmentally friendly cleaning list online so you know exactly what to buy.

- Have your home checked for asbestos. This is typically done before you move into the home. If you have a home that was built prior to the ban of asbestos, it is important to make sure there is none still lingering within the home.

- Stop using gas stoves in your home as well as certain types of space heaters. They release harmful chemicals that could be dangerous to human health.

- Have your home inspected for any mold, radon, or any other harmful chemical or bacteria that may be in your home. These types of inspections are traditionally done before you move in so keep that in mind as well.

- Use a good vacuum cleaner that has strong brushes to keep out chemicals and allergens that can accumulate in your home. Areas in your home which are most commonly visited must be cleaned thoroughly by using the vacuum several times.

- Most of the dirt comes in the home from the shoes. Keep a large mat out of every room that will reduce the amount of dirt, and other pollutants from getting into your home.

Smog

Smog is a mixture of air pollutants—nitrogen oxides and volatile organic compounds—that combine with sunlight to form ozone.

Ozone can be beneficial or harmful, good or bad, depending on its location. Ozone in the stratosphere, high above the Earth, acts as a barrier that protects human health and the environment from excessive amounts of solar ultraviolet radiation. This is the "good kind" of ozone.

On the other hand, ground-level ozone, trapped near the ground by heat inversions or other weather conditions, is what causes the respiratory distress and burning eyes associated with smog.

What Causes Smog?

Smog is produced by a set of complex photochemical reactions involving volatile organic compounds (VOCs), nitrogen oxides and sunlight, which form ground-level ozone.

Smog-forming pollutants come from many sources such as automobile exhaust, power plants, factories, and many consumer products, including paint, hairspray, charcoal starter fluid, chemical solvents, and even plastic popcorn packaging.

In typical urban areas, at least half of the smog precursors come from cars, buses, trucks, and boats.

Major smog occurrences often are linked to heavy motor vehicle traffic, high temperatures, sunshine, and calm winds. Weather and geography affect the location and severity of smog. Because temperature regulates the length of time it takes for smog to form, smog can occur more quickly and be more severe on a hot, sunny day.

When temperature inversions occur (that is, when warm air stays near the ground instead of rising) and the wind is calm, smog may remain trapped over a city for days. As traffic and other sources add more pollutants to the air, the smog gets worse. This situation occurs frequently in Salt Lake City, Utah.

Ironically, smog is often more severe farther away from the sources of pollution, because the chemical reactions that cause smog take place in the atmosphere while pollutants are drifting on the wind.

Where does Smog Occur?

Severe smog and ground-level ozone problems exist in many major cities around the world, from Mexico City to Beijing, and a recent, well-publicized event in Delhi, India. In the United States, smog affects much of California, from San Francisco to San Diego, the mid-Atlantic seaboard from Washington, DC, to southern Maine, and major cities in the South and Midwest.

To varying degrees, the majority of U.S. cities with populations of 250,000 or more have experienced problems with smog and ground-level ozone.

According to some studies, more than half of all U.S. residents live in areas where the smog is so bad that pollution levels routinely exceed safety standards set by the U.S. Environmental Protection Agency (EPA).

What are the Effects of Smog?

Smog is made up of a combination of air pollutants that can compromise human health, harm the environment, and even cause property damage.

Smog can cause or aggravate health problems such as asthma, emphysema, chronic bronchitis and other respiratory problems as well as eye irritation and reduced resistance to colds and lung infections.

The ozone in smog also inhibits plant growth and can cause widespread damage to crops and forests.

Who is most at Risk from Smog?

Anyone who engages in strenuous outdoor activity—from jogging to manual labor—may suffer smog-related health effects. Physical activity causes people to breathe faster and more deeply, exposing their lungs to more ozone and other pollutants. Four groups of people are particularly sensitive to ozone and other air pollutants in smog:

- Children: Active children run the highest risks from exposure to smog, as children spend a lot of time playing outside. As a group, children are also more prone to asthma—the most common chronic disease for children—and other respiratory ailments than adults.

- Adults who are active outdoors: Healthy adults of any age who exercise or work outdoors are considered at higher risk from smog.

- People with respiratory diseases: People with asthma or other chronic respiratory diseases are more sensitive and vulnerable to the effects of ozone. Typically, they will experience adverse effects sooner and at lower levels of exposure than those who are less sensitive.

- People with unusual susceptibility to ozone: Some otherwise healthy people are simply more sensitive to the pollutants in smog than other people and may experience more adverse health effects from exposure.

Elderly people are often warned to stay indoors on heavy smog days. Elderly people are probably not at increased risk of adverse health effects from smog because of their age. Like any other adults, however, elderly people will be at higher risk from exposure to smog if they already suffer from respiratory diseases, are active outdoors, or are unusually susceptible to ozone.

How can you Recognize or Detect Smog where you Live?

Smog is a visible form of air pollution that often appears as a thick haze. Look toward the horizon during daylight hours, and you can see how much smog is in the air. High concentrations of nitrogen oxides will often give the air a brownish tint.

In addition, most cities now measure the concentration of pollutants in the air and provide public reports—often published in newspapers and broadcast on local radio and television stations—when smog reaches potentially unsafe levels.

The EPA has developed the Air Quality Index (AQI) (formerly known as the Pollutant Standards Index) for reporting concentrations of ground-level ozone and other common air pollutants.

Air quality is measured by a nationwide monitoring system that records concentrations of ground-level ozone and several other air pollutants at more than a thousand locations across the United States. The EPA then interprets that data according to the standard AQI index, which ranges from zero to 500. The higher the AQI value for a specific pollutant, the greater the danger to public health and the environment.

Chapter 2

Causes of Air Pollution

Air pollution can be caused due to anthropogenic reasons such as waste incineration and smoking as well as natural reasons like volcanic eruptions, radon emissions and methane emissions. The chapter closely examines these key causes of air pollution to provide an extensive understanding of the subject.

Air pollution is caused by unwanted and harmful substances such as chemicals, dust, smoke emanating from the vehicles, suspended particles and other harmful gases in our environment. The increasing growth in the world population has led to over-exploitation of natural resources. Larger cities are becoming barren – devoid of vegetation and greenery – due to fast-paced industrialization. The population of these cities is increasing day by day; this has resulted in housing problems in cities. To solve this problem, people have built settlements (slums) where there is no proper arrangement of drainage, etc.

Smoke from industries, and use of chemicals in agriculture have aggravated air pollution. There have been several terrible accidents in factories.

The growth in the means of traffic is increasing, whether it is number of engines, buses, aircraft, scooters etc. The smoke coming out of these vehicles is continuously getting into the atmosphere, thereby polluting the atmosphere.

Deforestation has increased air pollution as the trees continually reduce pollution in the atmosphere. Plants absorb harmful pollution carbon dioxide for their food, and provide life-generating oxygen, but humans have indiscriminately cut them for residential and agricultural activities, and due to lack of green plants, there has been a decline in the natural process of purifying the atmosphere. Apart from this, nuclear particles from the nuclear test spread in the atmosphere which has a deadly effect on vegetation and animals.

Air pollution has a negative impact on the environment and life in general. The sources of air pollution can be divided into two parts:

Natural Sources

Pollutants polluting the air generated from natural sources are as follows:

- Dust blowing during storm.

- Smoke and carbon dioxide generated from fire in forests (Large amounts of smoke is produced by forest fires, which completely encircles the surrounding villages and cities and spread deadly pollution to humans and other living organisms).

- Methane gas emitted from decomposition substances in swamps.

- Carbon dioxide.

- Bacteria and viruses generated from wastes, etc.

- Carbon dioxide free from pollen of flowers.

- Cosmic dust generated due to collision of comets, asteroids and meteors, etc with the Earth.

- Volcanic eruptions.

- Evaporation of organic compounds and natural radioactivity.

- Erosion of rocks through air.

Human Sources

Humans are responsible for polluting the air through fossil fuels, agricultural activities, gases and smoke emanating from industries and vehicles, apart from mining operations and indoor pollution. Air pollution is mainly caused by combustion of fossil fuels such as petroleum substances, coal, wood, dry grass burning and construction activities. Motor vehicles produce excessive toxic gases such as carbon monoxide (CO) and hydrocarbons (HC) and nitrogen oxides (NO), which lead to air pollution.

Constructions of residential and commercial activities as well as road construction activities, etc. are also responsible for air pollution.

Air pollution caused by humans can be divided into the following processes:

- By combustion process,

- Combustion in domestic operations,

- Combustion in vehicles,

- Combustion for thermal electrical energy,

- By agricultural activities,

- By industrial constructions,

- By the use of solvents,

- By molecular energy related projects,

- Other reasons.

By Combustion Process

Generally, air pollution is of two types – indoor and outdoor air pollution. Energy is needed from cooking to construction of bricks, cement etc. The energy used for domestic operations is obtained from coal, wood, cooking gas, kerosene etc. The combustion of these fuels generates carbon dioxide, carbon monoxide, sulphur dioxide etc. and incomplete combustion of fuels produces many types of hydrocarbons and cyclic compounds. This type of combustion has two types of effects in the atmosphere. On one hand, these harmful gases pollute the air and on the other hand the amount of oxygen present in the air reduces, which is dangerous for life.

Energy is also required in the operation of vehicles and machines etc. This energy is obtained by combustion of various types of fuels. Among the outdoor causes of pollution, petrol or diesel are used as fuel for combustion in buses, cars, trucks, motorcycles, scooters, diesel, rails etc. There is a huge amount of black smoke coming out from them, which pollutes the air. The smoke that comes out of diesel vehicles contains hydrocarbons, nitrogen and sulphur oxide and micro-carbon compounds. Carbon monoxide and lead are present in gas fired vehicles. Lead is an air pollutant material.

According to an estimate, a motor vehicle spends the amount of oxygen in one minute which is equivalent to what 1135 people spend in breathing. Nitrogen oxides and nitrogen dioxide also arise from combustion of diesel and petrol in vehicles, which produces chemical smog by hydrocarbons in the sunlight. This smog is very dangerous for humans. In 1952, the city of London was surrounded by smog for five days, causing 4,000 people to die and millions of people who had become victims of cardiovascular disease and bronchitis.

As per an estimate, vehicles emit 60 tonnes of particulate matter, 630 tonnes of sulphur dioxide, 270 tonnes of nitrogen oxide, 2040 tonnes of carbon dioxide per day The ash that is produced on burning of coal, in the form of wastes, is dumped outside. This ash flies through the air and pollutes the atmosphere.

Sources of Air Pollution

There are different factors responsible for spreading air pollution in cities and rural areas:

Factors Responsible for Spreading Air Pollution in Cities

The biggest source of pollution in cities is polluted smoke emanating from vehicles and industrial establishments. Carbon monoxide emanating from air conditioners and vehicles is one of the major air pollutants in the cities. It is a poisonous, colourless gas, which is formed through burning fossil fuels such as coal, petroleum and natural gas.

Toxic air pollutants, such as sulphur dioxide, nitrogen oxides and carbon dioxide, etc. coming out of the factories, automobiles are major causes of air pollution. All the industries and manufacturing plants emit air pollution and hence they are significant contributors in generating acid rain in the cities.

Other sources of air pollution in the cities include dust and dirt produced by construction industry and the factories. Using chemicals for domestic cleaning and painting purposes also contribute to polluting indoor environment in the houses with inadequate ventilation. Already indoor air pollution is fast-spreading in the cities.

Factors Responsible for Air Pollution in Rural Areas

Although mostly urban factors are responsible for air pollution, but rural areas are also contributing to air pollution. Tractors being used in the villages for agriculture purposes are spreading air pollution and then dust flying in the field during farming is also playing a major role in polluting the air.

The reason for the pollution caused by volcanic eruptions is the volcanic ash produced on the surface of the earth and the large quantity of lava. Both natural and human causes have made the situation worst as far as air pollution is concerned.

Human Activities Responsible for Air Pollution

- Manufacturing Industry: Emissions from manufacturing industries are a major factor responsible for spreading air pollution. There are many harmful gas and solid particles in the smoke coming from the factories which enter the atmosphere and spread air pollution. Due to the continuous mixing of nitrogen, sulphur, carbon monoxide and carbon dioxide gases and other chemical waste in the air, air quality is getting worse everywhere.

- Smoke coming out of vehicles: With a large amount of smoke coming from vehicles, air pollution is spreading huge amounts of pollutants worldwide. The effects of air pollution due to vehicles can be clearly seen in every city. In vehicles, combustion of petroleum and other fossil fuels, resulting in toxic gases such as carbon dioxide and carbon monoxide release and pollute the air. Transport is an important part of our life which we cannot ignore. Running of cars, heavy trucks, trains, water vessels and aircraft necessitate combustion of fossil fuels, which emits large amounts of polluted fumes. Carbon monoxide, hydrocarbons and solid particles present in the smoke emitting from the vehicles are all dangerous air pollutants.

- Generating electricity: Coal and other fossil fuels are used extensively to run power plants. Due to the combustion of fossil fuels for generating electricity in these plants, there is a large scale production of various pollutants such as sulphur dioxide, carbon dioxide and nitrogen oxides that pollute the air.

- Emissions through chimneys: In the manufacturing plants, fossil fuels continue to be emitted by long chimneys. In this smoke, carbon monoxide, biological compounds and various chemical gases are found in the air and spread air pollution. The petroleum refinery industry also emits large amounts of hydrocarbons in the air, which is a dangerous pollutant.

- Earth-mining to remove ores: Activities such as continuous drilling, blasting, etc. are being carried out to remove ores and coal from various metals from the womb. In addition to these activities, transportation is also used for this purpose. Apart from methane, carbon monoxide, sulphur dioxide etc., mites, dust particles also pollute the air.

- Agricultural work: To increase the yield of crops, farmers use ammonia-based fertilizer, which is a harmful air pollutant. Apart from this, farmers also use many toxic insecticides in the fields to keep their crops safe. These insecticides emit many unwanted chemicals in the atmosphere, causing air pollution.

- Indoor air pollutants: Indoor air pollution is spread by many chemicals such as leads and paints used by humans in cleaning and coal, wood, cooking gas, oats, kerosene etc. in cooking. Chemicals are used in various functions at home or offices and where there is less ventilation, they can prove to be deadly. The solvents used in making polishing and spray paints on furniture are mostly flying hydrocarbons. When the furniture is polished or painted, then these hydrocarbons fly into the air.

- Molecular energy related projects: The isotopes used to generate atomic bombs and atomic electricity are temporary. At the time of the explosion, they spread far and wide in the atmosphere and later fall down on the earth as incubators, which leave their deadly effects.

The effect of the atom bombs dropped on Hiroshima and Nagasaki remained there for a long time.

- Bodies of animals: People carry dead animals from the settlements and take out the skins and leave the rest in the open. When these dead bodies rot, excessive foul smell emerges which causes air pollution.

- No Cleanliness of Toilets: The region's air is polluted by no regular cleaning of public and private toilets.

- Decomposition of garbage wastes and non-cleaning of drains: People often throw garbage in the street or out of the drains in their houses, which spread bad odour and the same also develops due to poor drainage facilities, which causes virus of various diseases and affects human health.

Gases Polluting the Air

Different types of gases which are major air pollutants include the following:

- Ozone Gas: Ozone is both good and bad. Harmful ozone gas is found in the lower part or near ground level of our atmosphere which causes asthma and other respiratory disorders. This gas is generated upon chemical reaction of pollutants emitted by vehicles, power plants, industrial boilers, refineries, chemical plants, and other sources in the presence of sunlight. But ozone layer at higher altitudes, located 6-30 miles from the surface of the biosphere, protects us from ultraviolet radiation.

- Sulphur Dioxide gas: Combustion of fossil fuels results in sulphur dioxide gas which is highly toxic and it is also responsible for acid rain.

- Nitrogen oxide gas: Nitrogen dioxide gas also has significant contribution in producing acid rain.

- Carbon monoxide gas: This poisonous gas, which is primarily emitted by the automobiles, is a dangerous air pollutant. Apart from the smoke coming out of the vehicles, carbon monoxide emissions are emitted by air conditioners, fridges and heaters inside the house.

- Particulate matter (PM): These pollutants, the sum of all solid and liquid particles suspended in air contained produce respiratory related problems by entering our lungs in the form of smoke of vehicles and factories.

Preventing the Causes of Air Pollution

The best way to control air pollution is to prevent air pollution factors. By regular checking of air quality, we get detailed information about a particular location of air pollutants and we can try to prevent them.

The factors of air pollution can be largely controlled by the development and use of green energy. That is why governments around the world are focusing on the development of green energy.

The use of solar and wind power has also proved to be effective in preventing air pollution. They pollute the air less than conventional energy sources.

Promoting the use of Public Transport

We should use more and more public transport modes. If we do all this, then the number of cars on the road will be reduced and air pollution can be controlled to a great extent.

Using Energy Resources Wisely

It is a bitter truth that many fossil fuels are burned to generate power, which causes large amounts of air pollution. Therefore, we should use fossil fuels intelligently and focus on the development and expansion of hydroelectric projects to reduce the quantity of contamination.

We need to develop a tendency to recycle and reuse things. Bulk pollution is spread by the manufacturing industry. If we recycle and reuse items like plastic bags, clothes, paper and bottle, then it can be helpful in reducing air pollution, as heavy pollution occurs due to combustion of fossil fuels in the production of any new item.

Anthropogenic Causes of Air Pollution

Human activity is responsible for most of the world's air pollution, both indoors and outdoors. Everything from smoking cigarettes to burning fossil fuels tarnishes the air you breathe and causes health problems as minor as a headache to as harmful as respiratory, lung and heart disease.

Man is at least partially at fault for most of the world's major air pollutants. Carbon dioxide is one of the most highly prevalent, comes from the combustion or burning of fossil fuels and other organic materials. Nitrogen oxide and dioxide, while both natural components of the Earth's atmosphere, occur in greater amounts due to human actions and are the cause of smog and acid rain.

Pollutants also include chlorofluorocarbons (CFCs), were widely used as refrigerants and aerosol propellants. These chemicals damage the ozone layer, which is why the Environmental Protection Agency banned them in 1978.

Particulates, microscopic particles of soot, pose yet another common hazard. Smoke from burning coal and diesel fuel has been one major source of particulate emissions. In addition to being harmful to breathe, particulates form a dark film on buildings and other structures.

Causes of Air Pollutants

The burning of fossil fuels such as coal and gasoline is the single largest source of air pollutants. Fossil fuels continue to be in wide use for heating, to operate transportation vehicles, in generating electricity, and in manufacturing and other industrial processes. Burning these fuels causes smog, acid rain and greenhouse gas emissions.

Burning fuels also increases some heavy metal contaminants and the amount of soot in the air. Power plants and factories emit much of the sulfuric air pollutants. In all, industrialized nations – particularly the United States and the Soviet Union – are responsible for most of the world's air pollutants.

Pollution Effects

Smog is one of the most dangerous air pollutants to humans and other biological organisms. It is made when coal and oil containing minor amounts of sulfur are burned. The oxides of these sulfur particles form sulfuric acid, which is toxic to life and damaging to many inorganic materials. Air pollution can damage human life, especially in major cities where there is a conglomerate of industries and fumes from vehicles.

Pollution harms the living environment. Sulfur dioxide, nitrogen oxides and peroxyacl nitrates enter leaf pores and damage plants that way. Pollutants also break away the waxy coating of leaves that prevent excessive water loss, causing further damage to crops and trees that are important to the surrounding environment.

Deadly Pollution Incidents

When man-made pollution aggregates over a city with a large population, dangerous situations can develop quickly. Two historical incidents of major pollution-related deaths and illnesses show how badly pollution can affect humans over a short period.

The first occurred in Donore, Pennsylvania, in 1948. Over several days, a high-pressure weather system trapped a large mass of stagnant air over the city, leading to dangerous levels of smog. The smoke from steel production had nowhere to go and accumulated in the air, causing 20 deaths and 6,000 cases of illness. In London, in 1952, a similar situation caused between 3,500 and 4,000 deaths in five days. While air pollution illnesses and deaths usually don't occur over such short periods of time, these are examples of worst-case scenarios with the possibility of occurring again if air pollution isn't mitigated.

Waste Incineration

Incineration is one of the waste treatment methods used to deal with tremendous amounts of human-made garbage created every single day. It is part of the thermal waste treatments, as it uses high temperatures. This process involves the combustion of organic substances contained in waste materials.

Incineration converts waste into ash, flue gas and heat. Ashes are mostly formed by the inorganic constituents of the waste and may take the form of solid lumps or particulates carried by the flue gas. The flue gases must be cleaned of gaseous and particulate pollutants before these are released into the atmosphere.

In the past, incineration was only conducted without separating materials thus causing harm to people nearby. This resulted in risk for plant workers and the environment. Most of such plants and incinerators never separated the most harmful pollutants nor generated electricity, and now-adays there are still many that don't do it.

As mentioned, in some incineration plants, the technique is used for generation of electric power, which is a good way to benefit from this process. However, there is still a lot of improvement to be made in the origin of the problem: generate less garbage or recycle correctly the humans' waste, especially transforming organic waste into composting.

The problem with waste is that, even planet earth is huge; we don't have enough land to fulfill with garbage. Even though incineration generates waste, the mass is reduced by approximately 95%, which makes the landfill problem way easier.

Other Waste-to-energy Techniques

Waste incineration is by far the most used process on the garbage treatment, but there are other techniques used to convert this trash into power energy:

- Gasification
- PDG
- Anaerobic digestion
- Pyrolysis

Types of Waste to Incinerate

Municipal Solid Waste

It is the solid portion of the waste (not classified as hazardous or toxic) generated by households, commercial establishments, public and private institutions, government agencies, and other sources. This waste stream includes food and yard wastes, and a multitude of durable and non-durable products and packagings.

Hazardous Waste

Hazardous waste is defined by EPA under the Resource Conservation and Recovery Act (RCRA) as a waste material that can be classified as potentially dangerous to human health or the environment on the basis of different criteria. This type of waste is found usually on manufacturers, service and wholesale-trade companies, universities, hospitals, government facilities, and households.

The hazardous waste can be classified as a threat for the following reasons:

- Waste of easy ignition.

- Corrosive waste for materials and/or people.

- Reactive waste: meaning it can explode, catch fire or give up harmful gases.

- Toxic waste.

Medical Waste Management

Usually biomedical, this type of waste can have infectious or toxic characteristics, so they need to be treated correctly to avoid creating a public health issue. The whole medical industry generates medical waste, but hospitals are the main medical waste generators, with almost 30 pounds per day per hospital on average.

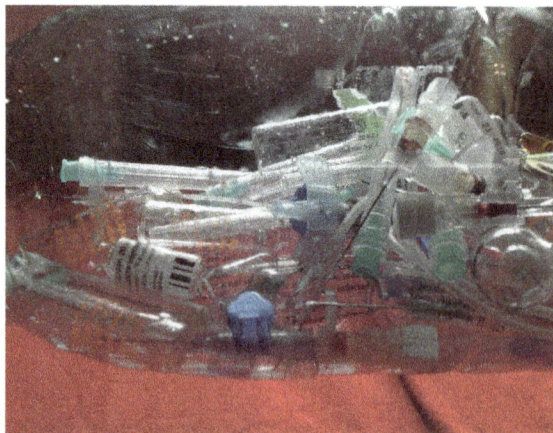

Waste Incineration Causes Air Pollution

Mass cannot be created nor destroyed, so the reduction of volume during the waste incineration process has its effects, and that is the emission of flue gas to the atmosphere. In other words, air pollution. These different types of emissions depend on the waste incinerated:

Furans and Dioxins

The emission of furans and dioxins is the biggest issue in the waste incineration process, giving that these are staidly injurious to health. Some governments have regulated these types of activities and obliged incineration plants to buy new machinery equipped with special equipment to clean emission of gases from these injurious components.

Carbon Dioxide

Carbon dioxide (CO_2), a major player in the global warming issue, is also produced in vast amounts during garbage incineration. These emissions are due to materials that include carbon in its composition, when incinerated, produce carbon dioxide.

Other Gases

A part from the major gases mentioned, a huge variety of other gases are emitted during the incineration of trash. On this large list, the most present volatiles are: sulfur dioxide, hydrochloric acid, fine particles and heavy metals.

Smoking

It might seem insignificant if you look just at one cigarette emissions, but take into account the quantity of cigarettes burned per day and it becomes a clear cause of air pollution. Recent studies in 2018 show that more than the 20% of the world's population smokes, which is an absolute value of more than 1 billion people smoking. And do you think they smoke only one cigarette per day?

Moreover, non-smokers are passively toxified by inhaling the fumes from smokers. The University of Minnesota estimates that up to 90 percent of the American population is routinely exposed to second-hand smoke, which means almost everyone is exposed to smokers' risks.

Tobacco smoke contains more than 50 carcinogens, making it an especially lethal form of air pollution. It emits a series of toxic chemicals including a series of organic and inorganic chemicals, some of which are carcinogenic.

Air Pollution

Growing Tobacco

Tobacco is a very fragile plant, so during massive crops a lot of pesticides, chemicals and herbicides are used. Methyl bromide is a chemical that stands out on the list due to its well-known effects on depleting the ozone layer.

Moreover, a lot of these pesticides and herbicides used when growing tobacco reach into the groundwater. An example is Aldicarb, a poisonous pollutant that can kill wildlife and humans, and was found in groundwater in 27 U.S. states in 2005.

Manufacturing

The manufacturing process of tobacco release several pollutants into the environment: ammonia, ethylene glycol, hydrochloric acid, hydrogen fluoride, methyl ethyl ketone, nicotine and nicotine salts, nitric acid, phosphoric acid, sulphuric acid, and toluene.

Supply always meets demand in every single product, and efforts should focus not on regulating this process, but on eradicating tobacco demand.

Transport

As any other product sold worldwide, major players grow and manufacture tobacco in few locations, and from there tobacco is spread worldwide using any mean of transport possible. Shipping vessels, trucks and planes, which are responsible for emitting tons of CO_2, are used to deliver tobacco everywhere. There is not a single village or urban mile where you cannot find it.

Consumption

There are 4.5 trillion filtered cigarettes smoked around the world every year, and all of those cigarette butts and fumes end up somewhere. Do we really think about it?

Smoking cigarettes releases more than 50 carcinogens and other toxins into the air we all breathe. This pollutes the air and harms human and animal life. Non-smokers are also exposed to second-hand smoke respiration, and fatal consequences are a lottery.

Cigarettes vs. Diesel Car Exhaust

The magazine Tobacco Control released a study which compared the air pollution in a closed garage for 30 minutes of a diesel car exhaust and three cigarettes combustion (smokers will know you can smoke three cigarettes in less than 30 minutes).

The results were astonishing, PM2.5 levels were 10 times greater in cigarettes than in the diesel car. This study raised the concerns of the real involvement that tobacco has in the air pollution issue, and is now considered a major source of air pollution responsible of environmental impact worldwide.

Smoking Health Risks

There are several short term and long term effects of smoking, and as you can imagine these are not beneficial, but dangerous effects. Overall, smoking has been one of greatest health hazards among humans, killing millions of people every year worldwide.

Non smokers can also suffer from these effects if they are exposed through passive smoking. In other word, breathing second hand smoke from smokers around them.

The most common disease associated to smoking is lung cancer, but this type of critical disease can be developed in multiple body parts:

- Mouth,
- Lips,
- Throat,
- Voice box (larynx),
- Oesophagus (the tube between the mouth and stomach),
- Bladder,
- Kidney,
- Liver,
- Stomach,
- Pancreas.

Smoking also damages your heart and your blood circulation, increasing your risk of developing conditions such as:

- Coronary heart disease,
- Heart attack,
- Stroke,
- Peripheral vascular disease (damaged blood vessels),
- Cerebrovascular disease (damaged arteries that supply blood to the brain).

Industrial Sources of Air Pollution

Fertilizer Complex

The fertilizer complex releases oxides of nitrogen and dust particles whose size ranges from sub-micron to 1000 microns. Dust particles may be evolved from the processes as drying, burning, calcining, grinding, screening, mixing, conveying or packaging. The major sources of dust emission can be listed in table.

Table: Possible source of dust emission in fertilizer industry.

S. No.	Product	Dust emitting particles
1	Urea	• Urea dust from prilling Tower.
2	Ammonia	• Coal and coke dust.
3	Ammonium sulphate	• Sulphur and pyrite dust. • Gypsum dust.
4	Calcium ammonium nitrate	• Lime Stone and product dust.
5	Phosphate Fertilizers	• Pyrite and sulphur • Rock Phosphate.
6	Mixed Fertilizer	• Product dust due in the granulation process.

The recorded annual average dust fall at Sindri Township.

Table: Dust fall at Sindri township (mg/sq. cm/month).

Total dust fall	1.25
Insoluble solid	0.93
Insoluble ash	0.62
Soluble Solid	0.33
Soluble Sulphate	0.03

The average dust fall in various cities of the world can be listed below:

Cement Factories

The cement dust is common air pollution around cement factories and construction sites. Chemically it is mixture of oxides of aluminium, potassium, silica, calcium and sodium.

The experimental results have shown that this dust reduces the size and number of leaves in plants. Growth reduction by cement dust has been reported by Bohne, Darley and Lerman. It has been found that dust affected plants in spite of their reduced growth are able to survive and complete their life cycle.

The studies revealed that there was a definite reduction of 33.34 in the number of grains per spike of dusted plants. It was also noted that weight and volume of 10000 grains was reduced by 11.4 and 12.6% respectively. The researches have revealed that the reduction in grains per spike was due to partial failure of pollen grain germination on dust-laden stigmas and eventually lack of fertilization in the ovary.

Hence from above data, it can be concluded that in areas polluted with cement dust, cultivation of plants in general and wheat in particular will suffer a sizable loss in terms of biological and economic field.

Thermal Power Stations

These stations are the principal sources of SO_2. It is best known and man-made pollutant. Fortunately because of its high reactivity and short life in atmosphere, no significant build up is reported. However, all industrialized areas suffer from SO_2 fumes.

Sulphuric Acid Industry and other Industries

The industries producing sulphuric acid either by contact process or by Lead Chamber process discharge large quantities of SO_2 in the air and pollute the atmosphere upto a few kilometer distances.

Many other industries also involve in one way or the other SO_2 causing a big air pollution in the region. The stack gas analysis data is given in table.

Table: Stack gas analysis data.

Name of plant	Pollutant	Ave. value	Min. value	Max. value
Sulphuric acid				
1. Plant	SO_2	0.19%	0.11%	0.30%
2. Plant	SO_2	0.45%	0.50%	0.57%
3. Plant	SO_3	0.06%	0.04	0.04%
4. Plant	SO_3	0.12%	-	-
5. Plant	$NO_2 SO_2$	0.36%, 0.40%	-	0.48%,0.50

Aeroplanes

The fuel in aeroplanes on combustion produces SO_2 as pollutant in the upper part of the atmosphere. This SO_2 forms sulphurous (H_2SO_3) and sulphuric acids — which are poisonous and fall on the ground with the rain and pollute the soil and rivers. In general, sulphuric acid destroys plants, animals, water bodies and also the fertility of the soil completely.

The main reactions are as follows:

$$SO_2 + H_2O \rightarrow H_2SO_3$$

$$H_2SO_3 + O \text{ (air)} \rightarrow H_2SO_4 \text{ Sulphuric acid}$$

$$2SO_2 + O_2 = 2SO_3$$

$$SO_3 + H_2O = H_2SO_4$$

Fluoride Industry

The industry produces fluoride compounds which have serious effects on plants, animals and human beings. The toxic effects of fluoride on live-stock arise from ingesting contaminated forage on which fluoride dust has been settled.

Cattle and sheep are very much susceptible to fluorine poisoning. Horses and poultry develop fluorisis due to large increase in fluoride content in their bones and teeth acquiring a stiff posture and becoming lame. In cattle the milk production decrease and they lose their weight.

Hydrogen fluoride destroys the leaves and fruits in the plants. The details of fluoride production and stack gas analysis data are presented in the tables below.

Name of Plant	Products	Production in Tons/day	Types of Emmission
Aluminium		5 Fluride (Alf$_3$)	HF, SO$_2$, AlF$_3$
Hydrogen fluoride	HF	15	HF, SiF$_4$, SO$_2$
Cryolite Plant	Cryolite	15	Cryolite HF
Sulphuric Acid Plant	20% oleum 90% Sulphuric Acid	55	SO$_3$, SO$_2$ acid mist

Table: Stack gas analysis data.

Name of plant	Pollutant	Avg. value	Max. Value
Hydrofluoric Acid Plant	HF	92mg/m³	170m/m³
Cryolite/AeF$_3$ plant	HF	238 mg/m³	420mg/m³

Table: Fluoride content of Jowar leaves.

Sample Particular	Average fluoride dust deposit over the leaf surface (mg of HF/gm)	Average fluoride content in tissues of leaf (mg of HF/gm)
Jowar leaves near the factory.	0.945	0.383
Jowar leaves 4 kms away from factory.	0.031	0.156

Nitric Acid Plants

Nitric acid plants generally release oxides of nitrogen in the atmosphere which are serious pollutants. There is an increasing concern over their levels in the atmosphere on account of their role in photo-chemical reactions which result in the formation of highly toxic substances in polluted environments.

These oxides of nitrogen also consume a lot of ozone of the above atmosphere.

The probable reactions are as follows:

$$NO + O_3 \rightarrow NO_2 + O_2$$
$$NO + O_3 \rightarrow NO_2 + O_2$$
$$NO_2{}^* \rightarrow NO_2 + Hv$$

Chloralkali Plants

Emission from chloralkali plants includes:

- Chlorine gas
- Carbon monoxide
- Carbon dioxide and
- Hydrogen.

Iron and Steel Industry

The emission from iron and steel industry includes:

- SO_2
- Metal oxides and
- CO and CO_2

Radioactive Natural Sources

The radioactive elements which occur in igneous rocks and soils can be divided into three parts:

- Uranium series (U^{238})
- Thorium series (Th^{232}) and
- Actinium (U^{235})

The gases such as radon and thoron which give rise to a series of radioactive decay products which get deposited on SPM in the atmosphere. So samples of suspended particulates collected by high volume samplers exhibit definite amounts of absorbed radioactivity.

Incineration of Refuse or Solid Waste

The refuse or solid wastes contain the following pollutants:

- Benzo (a) pyrene
- Organic acids
- Particulate
- Ammonia
- Oxides of sulphur
- Hydrocarbons
- Aldehydes
- Carbon monoxide and
- Oxides of nitrogen

Industrial Sources of Odour Pollution

Offensive smell is the most strident kind of pollution and its reduction involves not only chemical, physical and engineering problems but also social, physiological and psychological.

Generally tanneries, meat and fish processing, tallow melting and refining and gut cleaners create odour pollution in the area in which they are located. Tallow is the term, which describes animal fat from all sources. Its main constituents are palmitates, $C_{15}H_{31}COOR$ and sterates $C_{17}H_{35}COOR$ of glycol, $C_3H_8O_3$ with some oleatets, $C_{17}H_{33}COOR$ of glycerol.

The melting point of tallow is about 45°C. Below 45°C, there is hardly any smell. The blue fumes which come out due to heating of fat above 45°C are due to formation of acrolein and other burnt and carbonized products which are irritant and offensive.

In guts, the gaseous decomposition products are released producing a serious pollution of air.

Cooking smells, the gases from sewage and domestic refuse, maggot (worm like larva) breeding and catering industry also cause air pollution.

The main pollutants from kitchens are:

- Aldehydes
- Hydro-carbons
- Carbon monoxide
- Carbon dioxide and
- Obnoxious gases.

Emissions

The coarse particles of solid particulate matter which settle out of the atmosphere by gravity are called emissions. These emissions pollute the atmosphere.

SPM (Suspended Particulate Matter)

These are particles, either liquid or solid dispersed in gaseous medium. The solid particles may be irregular may be irregular in shapes but liquid particles are sphericity in shape. Particles larger than 100 µm tend to settle out of the air by gravity. They cause air pollution.

Sulphur Dioxide

The coal burning power plants release 41% of the total emission of SO_2. The combustion of coal and petroleum products together liberates about 18 million tons of SO_2 to the atmosphere from vehicles.

Nitrogen Oxides

The fuel combustion sources produce 10 million tons of nitrogen oxides, with about 25% from electricity generation, 17% from industrial, 9% from residential, 46% from transportation sources and 3% from commercial operations.

Carbon Monoxide

99% CO comes from vehicles. In 1984, 213 million tons was discharged into the atmosphere.

Viable Particles

A large number of viable particles are found in the atmosphere over land during the growing season. These consist of pollens, microorganisms and insects. In fact, the pollens are aeroallergens and are found in the atmosphere due to grasses, trees, plants, bushes and weeds. The pollen grains produce hay fever and various types of allergic reactions in human beings.

The most important viable particles are yeasts, algae, fungi, spores, bacteria, rusts and moulds etc. Except algae, all other microorganisms can be transported by wind. They are generally responsible for infection in animals, plants and human beings.

The bacteria are found in air, water, soil, vegetation, food, dust particles and in the bodies of human beings and animals. They are always present in the atmosphere irrespective of climate cold, hot or rainy season.

Atmospheric Reactions

The oxides of nitrogen discharged into the atmosphere from combustion of fuels and industries are highly reactive.

They react with moisture or water and from various acids as follows:

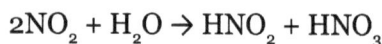

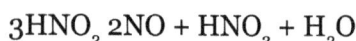

$$2NO_2 + H_2O \rightarrow HNO_2 + HNO_3$$

$$3HNO_2 \, 2NO + HNO_3 + H_2O$$

Similarly, the SO_2, which is discharged into the atmosphere from iron works, factories etc. produces the following products with H_2O and fall on the earth:

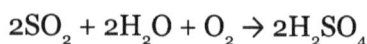

$$2SO_2 + O_2 \rightarrow 2SO_3$$

$$SO_3 + H_2O \rightarrow H_2SO_4$$

$$2SO_2 + 2H_2O + O_2 \rightarrow 2H_2SO_4$$

About 80 million tons SO_2 is released into atmosphere per year. The rate of H_2S formation is about 300 million tons per year.

Now take the case of CO_2. It is released by all forms of life during respiration and is assimilated by green plants during photosynthesis.

CO_2 which is released by carbonaceous fuels etc., has increased its concentration in the atmosphere. Emission of CO to the atmosphere is more than 200 million tons per year.

In fact, the chemical reactions occurring in the atmosphere convert gases or vapours into solid and liquid products by oxidation, reduction, condensation, polymerization, combination and other mechanisms. The photochemical reactions in the upper atmosphere convert complex molecules into simple molecules due to solar radiation.

Man-Made Sources of Air Pollution

Man is the main culprit for producing pollution in the air through use of coal, oil and natural gas as fuel and exhaust gases from automatic vehicles. Besides this, the industrial activities such as oil refining, iron and steel manufacturing, non-ferrous manufacture of rubber tyres, tubes and pulp and paper and other processes, industries have added new dimensions to the field of air pollution.

The man-made pollutants are listed in table below:

Gases and others	Tons per Year
H_2S	300 million
SO_2	80 million
CO_2	200 million
Organic sulphides and Mercaptans	50 million
NO_2	150 million

Mineral Acids

Nitric Acid Manufacture

It has been estimated that for the production of 1 ton of acid, about 30-100 pound of nitrogen oxides is produced, out of which about 70-80% of brown coloured nitrogen dioxide (NO_2) can be reduced to colourless nitric oxide (NO) by a catalytic reaction unit. Because of extra fuel requirement, about 25% of total nitrogen oxide is reduced to nitrogen and oxygen by catalytic reduction.

Main causes of air pollution.

Sulphuric Acid Manufacture

In this plant the main emissions are (SO_2) and sulphuric acid (H_2SO_4). It has been estimated that for the production of 1 ton of acid, the emissions of SO_2 range from about 25-70 pounds. Without acid must eliminators can reduce SO_2 to 0.02 from 0.2 pounds of acid must/ton of acid produced.

Emissions:

- SO_2
- SO_3

Hydrochloric acid manufacture

The principal emissions are:

- HCl,

- Chlorinated products,

- Unreacted hydrocarbons and

- Chlorine.

The contaminants can be minimized if the effective absorption tower or scrubbing system is used to remove residual HCl.

Phosphoric and Manufacture

In the electric furnace process-phosphate rock siliceous flux and coke are heated in an electric furnace to produce phosphorus.

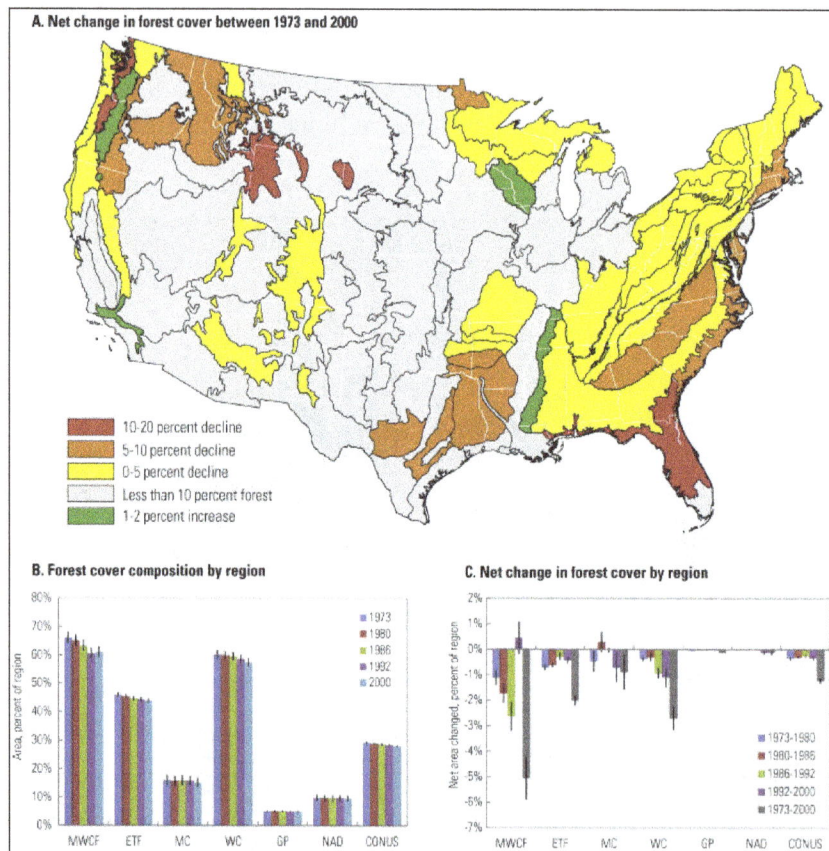

A. Net change in forest cover between 1973 and 2000

10-20 percent decline
5-10 percent decline
0-5 percent decline
Less than 10 percent forest
1-2 percent increase

B. Forest cover composition by region

C. Net change in forest cover by region

The principal emissions are:

- Phosphoric acid,

- Phosphorous Penta oxide (P_2O_5). Emissions of P_2O_5 are in the range of 3.6 to 5 pounds of elemental phosphorous (P_4) burnt.

In the Wet Process

H_2SO_4 and phosphate rock are reacted to form phosphoric acid and gypsum. Main emissions are: (a) gaseous fluorides with Some H, (b) P_2O_5 along with phosphoric acid.

Petrochemical Refineries

An oil refinery consists of:

- Distillation towers,
- Cracking towers,
- Pressure vessels,
- Heat exchangers,
- Valves,
- Pumps,
- Storage tanks.

From the refineries, we get the following emissions:

- Particulate,
- CO,
- HCHO,
- NH_3,
- NO_2,
- Gasoline,
- Kerosene,
- Fuel oil etc.

Emissions from Fuels, Incinerators and Internal Combustion Engine

When coal is used as a fuel, we get the following emissions:

- Particulate matter — 30-150 lbs/ton of fuel.
- SO_2 -60-120 lbs/ton of fuel.
- Other organics—20-30 lbs/ton of fuel.

The contaminants discharged to the atmosphere from fuel burning, incineration of refuse and internal combustion engine are as follows:

- Particulate matter (Soot and fly ash),

- Sulphur dioxide,

- Natural gas.

Exhausts from engine include:

- CO,

- Oxides of nitrogen,

- Hydrocarbons,

- Oxygenates of hydrocarbons,

- Lead compounds,

- Carbon particles,

- Motor oil,

- Non-volatile reaction products formed from motor oil in the combustion zone. High molecular weight olefins and carbonyl compounds are produced in the reaction.

The amount of above products depends upon the composition of fuel, method of firing and other factors.

Table: Emissions factors for uncontrolled automobile exhaust.

Types of emission	Emission ponds per 100 vehicles miles	Pounds per vehicle day
Particulates	0.8	0.022
Organic acids (Accetic)	0.3	0.007
Oxides of sulphur (SO_x)	0.6	0.016
Oxides to Nitrogen (NO_x)	8.5	0.202
Hydrocarbons (C_xH_y)	12.5	0.363
Carbon Monoxide	165.0	4.160
Aldehydes (RCHO)	0.3	0.007

Table: Emissions factors for diesel engines (pounds per 100 gallons of diesel fuel).

Types of emission	Emission factor
Particulates	110
Organic acids (Accetic)	31
Oxides of sulphur (SO_x)	40
Oxides to Nitrogen (NO_x)	222
Hydrocarbons (C_xH_y)	136
Carbon Monoxide (CO)	60
Aldehydes (RCHO)	10

Table: Emission from power plant (Ib/hr).

	Particulate	SO_2	NO_2
Coal	117.300	17.400	18.340
Oil	1.420	22.300	14.800
Gas	303	8	1.890

Natural Gas

The emissions from natural gas combustion are given in the following tables.

Table: Emission Factor for Natural Gas Combustion (pounds per million cubic feet of natural gas burnet).

Pollutant	Types of unit		
	Power plant	Industrial Process boilers	Domestic and commercial heating units
Particulates	15	18	19
Other organics	3	5	Nil
Oxides of sulphur (SO_x)	0.4	0.4	0.4
Oxides to Nitrogen (NO_x)	390	214	116
Hydrocarbons (C_xH_y)	Nil	Nil	Nil
Carbon Monoxide (CO)	Nil	0.4	0.4
Aldehydes (RCHO)	1	2	Nil

Combustion of Coal

In the combustion of coal, we get the following products:

- Carbon monoxide (CO),
- Oxides of nitrogen (NO_x),
- Sulphur dioxide (SO_2),
- Aldehydes (RCHO) and
- Hydro-carbons (C_xH_y).

The quantities of these pollutants depend on the composition of Coal, method of firing, the size of unit, atmospheric conditions, etc.

The Burning of Coal also Produces

- Fly ash emission,
- Particulate emission: The important constituent of particulate emission is benzo (a) pyrene which has been found to be carcinogenic to animals.

Combustion of Fuel Oil

The main pollutants are:

- Particulate emissions,

- Oxides of sulphur.

Kraft Pulp Manufacture

From the Kraft pulp industry, we get the following gaseous pollutants:

- H_2S,

- Marcaptans,

- Various other sulphides.

The main pollutants and their major sources of generation are presented in table.

Table: Main pollutant species and their major sources of generation.

Pollutant species	Major sources
Sulphur dioxide	Electricity generation oil refineries iron and steel works.
Smoke, dust, grit	Iron and steel works, generating station, foundries, gas works cement works.
Carbon monoxide	Combustion of fossil fuels.
Carbon dioxide	Combustion of fossil fuels.
Oxides of nitrogen	Nitric acid works, electricity generation, iron and steel works, fertilizer plant.
Ammonia	Ammonia works.
Sulphur trioxide	Sulphuric acid works, brick works.
Sulphides and sulphur	Generating stations, metal smelting. Rubber vulcanizing, coke ovens.
Chlorine and hydrogen	Chlorine works, secondary.
Chloride	Aluminium works, chromium works.
Chlorinated hydrocarbons	Dry cleaning works.
Mercaptans	Oil refineries, coke ovens.
Zinc oxide	Copper works.
Zn and Cd	Zinc industries.
Freon	Refrigeration works.
Zn, SO_4-Cd, and other	Rubber industries.
Cr, Ni, Co, etc.	Dyeing and Printing industries.

Noxious or Offensive Gases

The requirement of the 1906 Act of U.S.A is that the best practicable means shall be used to prevent the escape of noxious or offensive gases, whether directly or indirectly, into the atmosphere and for expression 'noxious or offensive gas' includes the list of gases and fumes given in section

27 (1) of the act of U.S.A. as extended by the 1966 and 1971 orders, and as effectively extended by section 11 (2) of the Clean Air Act 1958, it also includes smoke, grit and dust.

Agricultural Impact on Air Quality

While power and transportation industries are largely responsible for the majority of greenhouse gas emission in the United States, the US Environmental Protection Agency (USEPA) reported that agricultural activities accounted for about 12 percent of emission in 2012.

Major Sources of Greenhouse Gas Emission

There are four main agricultural activities linked to the production of greenhouse gases. These include soil management, enteric fermentation, manure management and fossil fuel consumption.

Main Pollutants from Agricultural Activity

Methane (CH_4) and nitrous oxide (N_2O) were the two main gases emitted by agricultural activity in 2012, Methane was primarily produced from enteric fermentation and manure management, while soil management, such as fertilization, was the largest source of nitrous oxide.

Additional Air Pollution from Agriculture

In addition to methane and nitrous oxide, agricultural activities have been linked to the emission of other dangerous gases and pollutants. These include carbon dioxide, ammonia, hydrogen sulfide, and airborne particulate matter, which has been linked to health problems.

How to Decrease Pollution Emission

Various methods to decrease air pollution have been identified and continue to be researched. Techniques including better manure storage, precision nutrient application, and air-breaks between farms can all help decrease the effect of agricultural practices on air quality.

Improving Nutrient Management

One way farmers can reduce nitrous oxide emission is to integrate the use of variable-rate fertilizer or manure management. This method encourages more efficient use of fertilizer by applying a unique amount to each field based on needs of that zone. Farmers use variable-rate manure application to apply varying amounts of fertilizer throughout the field depending on different zones found naturally within it. These zones are created based on soil tests.

In addition to reducing nitrous oxide emission, precision management can reduce production costs and decrease the risk of water contamination.

Manure Management

While strategies to control odor and dust, such as land application, have been widely adopted, methods to control gas emission still need focus.

Methane is the main gas produced by manure as it decomposes. When stored as a liquid in a lagoon or tank, manure decomposes anaerobically, creating methane emissions. However, manure decomposes aerobically when deposited naturally, creating little methane emissions. Because of this, handling manure as a solid versus a liquid can help decrease emissions.

In addition, storing manure in anaerobic containment areas that utilize technology to capture the methane produced as an energy source is a viable option. Not only does this method reduce emissions, but can also offset costs by decreasing the use of fossil fuel energy.

Livestock Management

Another way that methane enters the atmosphere is through enteric fermentation. Although this is a normal part of the digestive process, ruminant animals like cattle have a unique digestive system that causes them to be a major emitter of methane. Feed intake and quality influence this level of emission, with lower feed quality and higher intake causing higher methane emissions.

Natural Causes of Air Pollution

Whilst man-made pollution and poor air quality is major environmental concern, there are many natural sources of pollution which are often much greater than their man-made counterparts.

Natural sources of sulphur dioxide include release from volcanoes, biological decay and forest fires. Actual amounts released from natural sources in the world are difficult to quantify. In 1983 the United Nations Environment Programme estimated a figure of between 80 million and 288 million tonnes of sulphur oxides per year (compared to around 69 million tonnes from human sources world-wide).

Natural sources of nitrogen oxides include volcanoes, oceans, biological decay and lightning strikes. Estimates range between 20 million and 90 million tonnes per year nitrogen oxides released from natural sources (compared to around 24 million tonnes from human sources worldwide).

Ozone is a secondary photochemical pollutant formed near ground level as a result of chemical reactions taking place in sunlight. About 10 to 15% of low level ozone, however, is transported from the upper atmosphere (called the stratosphere), where it is formed by the action of ultraviolet (UV) radiation on oxygen (the ozone layer).

Natural sources of particulate matter are less important than man-made sources. These include volcanoes and dust storms. However, such sources do account for intense high particulate pollution episodes, occurring over relatively short times scales. It is not unknown for Saharan dust to be deposited in the UK after being blown thousands of miles.

Volatile organic compounds (VOCs) are naturally produced by plants and trees. Isoprene is a common VOC emitted by vegetation, and some believe it to be a more significant trigger for asthma an other allergic reactions than man-made irritants. Plant, grass and trees are also a source of pollen, which can act as triggers in some asthmatics. Pollen is in the air year-round, but the concentration is highest during the growing season, from March to the first frosts in autumn.

Natural pollutants found indoors include the dustmite, mould spores and radon gas.

Volcanic Eruptions

A volcano is an open fissure on the surface of the earth. Active volcanoes are those from which lava, volcanic ashes, rocks, dust and gas compounds escape on a regular basis (10.000 years are considered regular with volcanoes, so you can feel safe if you have one around) due to the phenomenon of volcanic eruptions.

In the world there are several active volcanoes which cause air pollution, danger to life forms and massive destruction of the land and the environment. Indonesia is the country with most active volcanoes in the world with 76 of them and total of 147 volcanoes.

Volcanic Eruptions release massive quantities of solid pollutants and gases, forming enormous clouds that can affect areas miles away from the volcanic eruption. Therefore, volcanoes are an international form of air pollution, but not just for us, as a lot of this greenhouse gases and aerosols go directly into the atmosphere.

Every volcanic eruption is different on impact, and therefore different on the quantity and a variety of pollutants emitted. On average the outgassed composition release is 79% water vapor (H_2O), 11,6% carbon dioxide (CO_2), 6,5% sulphur dioxide (SO_2) and 2,9% of other pollutants.

However, the range of pollutants released on a volcanic eruption include: carbon dioxide (CO_2), sulphur dioxide (SO_2), hydrogen sulfide (H_2S), hydrogen (H_2), hydrogen fluoride (HF), hydrogen chloride (HCl), bromide oxide (BrO) and carbon monoxide (CO). Highly exposure to these gases has detrimental impact on living organisms both terrestrial and marine.

On the other side, particulates are another source of air pollution produced by volcanic eruptions. Mainly ashes and including all types of sizes, the ones that help forming toxic clouds are usually PM10, PM2.5, PM0.3 and thinner.

These pollution clouds can travel major distances, like crossing oceans, dangerously affecting people and environments who didn't even notice the eruption. If you are facing a polluted cloud caused by a volcanic eruption, you may want to have a look at the following content on what to do and how to prevent these events.

Radon Emissions

Radon (which has the symbol Rn and its atomic number is 86) is a chemical element belonging to noble gases, very radioactive, colorless, odorless and tasteless. This is the reason why radon is also called invisible gas or "the silent killer". In solid form it is reddish. It is the earth's only naturally produced radioactive gas and comes from the breakdown of radium, uranium in soil, rock, and water.

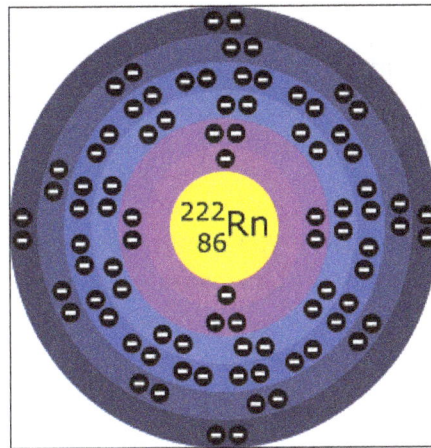

This air pollutant is highly radioactive in nature, and it can cause some serious health damages to people who breathe it. In fact, it is the second largest contributing factor to lung cancer in human beings after smoking. It causes an estimated 1,100 deaths from lung cancer every year.

Because the level of radioactivity is directly related to the number and type of radioactive atoms present, radon and all other radioactive atoms are measured in picocuries.

Radon Decay

Radon is diffused out of the air all the time in variable quantities depending on the pressure drop. Such pressure drops can accompany or precede the shearing of rocks in an earthquake. It disperses and decays very quickly, with a half-life of 3.8 days.

Radon decay products (RDPs) such as polonium (218), lead (214) and bismuth (214) are measured in working levels (WL).

Radon at Home

Why do some houses have high levels of indoor radon while nearby houses do not? The reasons lie primarily in the geology of radon, the factors that govern the occurrence of uranium, the formation of radon and the movement of radon, soil gas, and groundwater.

It can seep into buildings through cracks and holes in its foundations, where it can build up to dangerous levels. In Britain in 2018, the number of homes designated at risk was increased five-fold (from 100,000 to between 500,000 and 600,000), rendering millions more people officially vulnerable.

If you smoke or someone at home smokes, your family is exposed to radon, and therefore probabilities of lung cancer increase. Moreover, the only thing you can do to eradicate this type of gas is quit, and any time is good to reduce its exposure.

Due to health risks of radon exposure, it is recommended testing radon concentration before buying a house or while building it. However, if you want to analyze indoor air quality of your current home there is easy and fast techniques to test radon emissions by your own, for example with charcoal or digital tests.

Methane Emissions

Methane is a powerful greenhouse gas emitted by human activities such as leakage from natural gas systems and the raising of livestock, as well as by natural sources such as wetlands. It has a direct influence on climate, but also a number of indirect effects on human health, crop yields and the quality and productivity of vegetation through its role as an important precursor to the formation of tropospheric ozone.

Methane is a short-lived climate pollutant with an atmospheric lifetime of around 12 years. While its lifetime in the atmosphere is much shorter than carbon dioxide (CO_2), it is much more efficient at trapping radiation. Per unit of mass, the impact of methane on climate change over 20 years is 84 times greater than CO_2; over a 100-year period it is 28 times greater.

Multiple studies have demonstrated that a selection of measures to cut methane emissions can reduce near-term warming of the climate, increase crop yields and prevent premature deaths.

40%	60%	Years 12	84x
Agriculture is the key emitting sector of methane emissions, responsible for about 40%.	Globally, over 60% of total methane emissions come from human activities.	Methane remains in the atmosphere for about 12 years.	Methane warms the planet 84 times as much as carbon dioxide over a 20-year period.

Primary Sources of Methane Emissions

Atmospheric methane concentrations have grown as a result of human activities related to agriculture, including rice cultivation and ruminant livestock; coal mining; oil and gas production and distribution; biomass burning; and municipal waste landfilling Emissions are projected to continue to increase by 2030 unless immediate action is taken.

In agriculture, rapid and large scale implementation of improved livestock feeding strategies can reduce of 20% of global methane emissions by 2030, while full implementation of intermittent aeration of continually flooded rice paddies (known as alternate wetting and drying cultivation) could reduce emission from rice production by over 30%.

Emissions from coal mining and the oil and gas sector could be reduced by over 65% by preventing gas leakage during transmission and distribution, recovering and using gas at the production stage, and by pre-mine degasification and recovery of methane during coal mining.

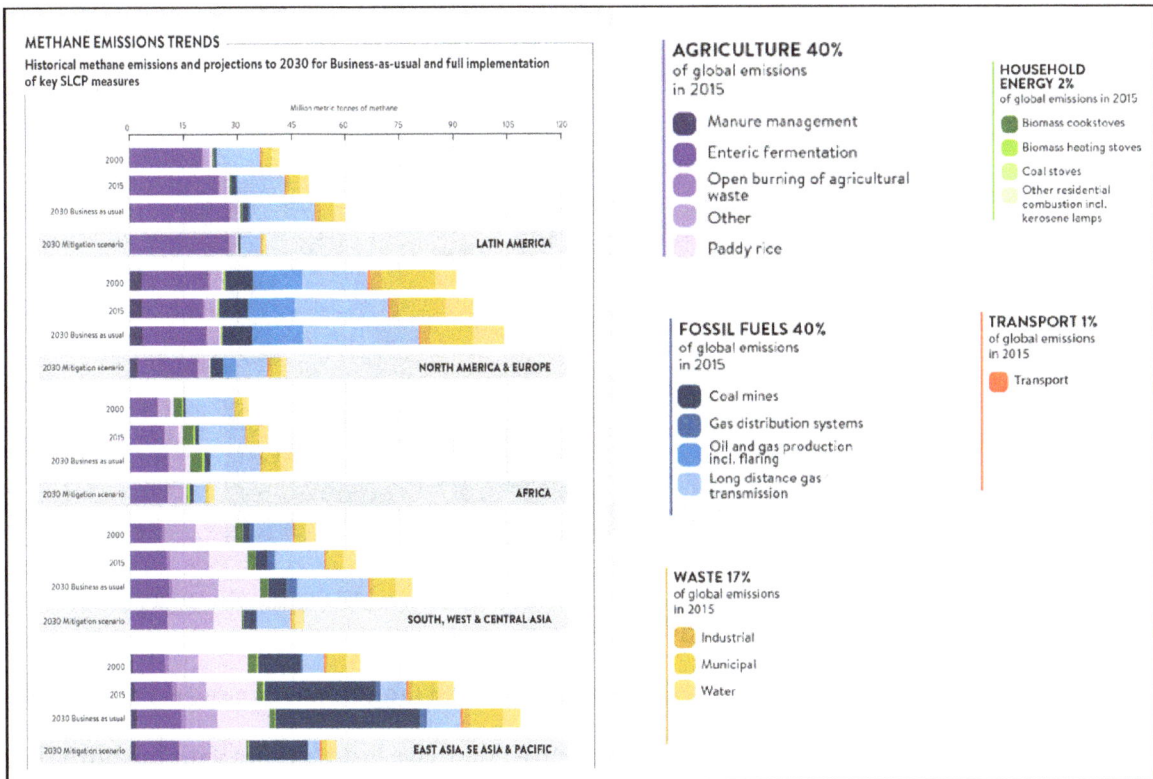

METHANE EMISSIONS TRENDS
Historical methane emissions and projections to 2030 for Business-as-usual and full implementation of key SLCP measures

AGRICULTURE 40% of global emissions in 2015
- Manure management
- Enteric fermentation
- Open burning of agricultural waste
- Other
- Paddy rice

FOSSIL FUELS 40% of global emissions in 2015
- Coal mines
- Gas distribution systems
- Oil and gas production incl. flaring
- Long distance gas transmission

WASTE 17% of global emissions in 2015
- Industrial
- Municipal
- Water

HOUSEHOLD ENERGY 2% of global emissions in 2015
- Biomass cookstoves
- Biomass heating stoves
- Coal stoves
- Other residential combustion incl. kerosene lamps

TRANSPORT 1% of global emissions in 2015
- Transport

Methane Impacts

Climate Impacts

Methane is generally considered second to carbon dioxide in its importance to climate change. The presence of methane in the atmosphere can also affect the abundance of other greenhouse gases, such as tropospheric ozone, water vapor and carbon dioxide.

Recent research suggests that the contribution of methane emissions to global warming is 25% higher than previous estimates.

Health Impacts

Methane is a key precursor gas of the harmful air pollutant, tropospheric ozone. Globally, increased methane emissions are responsible for half of the observed rise in tropospheric ozone levels.

While methane does not cause direct harm to human health or crop production, ozone is responsible for about 1 million premature respiratory deaths globally.

Forest Fire

Forest fire originated by natural causes, or in other words without human activity involved, are also known as bushfires or wildfires. These are given when fire has place in highly dense vegetation areas.

Furthermore, bushfires also contribute to deforestation, what indirectly affects air quality in forests and jungles that are considered Earth's lungs.

Australia is well known as a country with plenty of vegetation and wildlife. However, the country has an important issue with bushfires, as they cost more than £180 million each year (both natural and man-made).

Unfortunately, there are a huge amount of notable bushfires in history. Some examples of most important wildfires worldwide have been:

- China (1987) burning over 72.000km2 and Indonesia (1997) burning over 97.000km2.

- The U.S. and Canada have been the most active countries last 10 years with bushfires burning around 6.6 and 6.2 acres/year respectively.

A long time ago, palaeowildfires burnt plant material leading to fossil charcoal formation. Nowadays, fossil charcoal is being used as an indicator of palaeoclimatology in order to study evolution on bushfires over the years.

Causes of Bushfires

There are several different causes that lead to forest fire, but these are made more likely when the weather is hot and dry. Fallen leaves, dry grass or branches easily light up and can cause serious trouble to both urban and rural areas.

These terrible events often pass in just a few minutes, but they can last days, or in very unfortunate situations even months. The real threat when it comes to bushfires is high winds because these fan the flames and spread the blaze.

Forest Fire Pollution

Forest fire release pollutants like smoke, sulphur dioxide, nitrogen dioxide, ozone, particulate matter and carbon monoxide into the atmosphere. The bigger the bushfire, the bigger emissions.

In small amounts, carbon monoxide is present human bodies. The problem is that when inhaled in large amounts, and large amount of this gas are emitted on bushfires, it becomes toxic and can cause death.

Particulates on forest fire are either solid or liquid and consist of soot, tars, and other volatile organic substances. These particles can have different sizes: PM10, PM2.5, PM0.3 and smaller. They can cause cardiovascular and respiratory problems if they penetrate deeply into humans lungs.

Nitrogen oxides and sulphur dioxides are minor problems when it comes to forest fire. Nitrogen oxides are only released in big bushfires, as they only appear at greater temperatures than 1,500 degree centigrade. On the other hand, except when peat and muck soil are involved, sulphur dioxide only appears to be less than 0,2%.

Health Effects of Wildfires

Wildfire in California during the summer 2018 caused hazardous air conditions across the state, prompting air quality alerts and forcing many residents to take refuge indoors to avoid unhealthy exposure to bad air.

It is not just about foggy and hazy skies, human' risks and exposure are real. The air pollution from wildfires includes huge percentage of particles PM10, PM2.5, PM0.3 and thinner.

These health effects of forest fires include coughing, sore throats, extreme wheezing, cardiovascular illnesses and problems into lungs and bloodstream. Of course, sensible groups like children or elderly, are more likely to acquire these symptoms.

However, what do you have to do in the event of a forest fire? Firstly, you should be worried about the dangers of wildfires, so stay tuned to a local station for recommendations. Additionally, to protect yourself from air pollution: wear some kind of pollution mask, avoid staying outdoors and close all the opening from your house to prevent air pollution to enter. Finally, avoiding riding cars will help reduce SMOG, as this is produced with the mixing of combustion gases and the air pollution produced in forest fires.

References

- Causes-and-sources-of-air-pollution, environmental-issues: indiacelebrating.com, Retrieved 21 June, 2019

- Manmade-causes-air-pollution-8674978: sciencing.com, Retrieved 2 April, 2019

- Waste-incineration, anthropogenic, causes, air-pollution: airgo2.com, Retrieved 21 May, 2019

- Smoking, anthropogenic, causes, air-pollution: airgo2.com, Retrieved 26 July, 2019

- 28-main-sources-of-air-pollution-by-industries, air-pollution: environmentalpollution.in, Retrieved 5 February, 2019

- Natural-air-pollution, air-quality: enviropedia.org.uk, Retrieved 19 April, 2019

- Volcanic-eruptions, natural, causes, air-pollution: airgo2.com, Retrieved 17 May, 2019

- Methane, slcps: ccacoalition.org, Retrieved 14 July, 2019

- Forest-fire, natural, causes, air-pollution: airgo2.com, Retrieved 31 March, 2019

Chapter 3

Types of Air Pollutants

There are various materials which are classified as air pollutants such as nitrogen oxides, ozone, sulfur dioxide, carbon monoxide, volatile organic compounds, mercury, ammonia, lead and chlorofluorocarbons. The topics elaborated in this chapter will help in gaining a better perspective about these air pollutants as well as short-lived climate pollutants.

Air Pollutant is any pollutant agent or combination of such agents, including any physical, chemical, biological, radioactive substance or matter which is emitted into or otherwise enters the ambient air and can, in high enough concentrations, harm humans, animals, vegetation or material.

These pollutants can have a terrible impact on the health of anyone who is exposed, which means every single person in the entire country. The pollutants can also protract a horrible affectation on the environment, and can cause property damage.

Of the six pollutants named in the previous paragraph, particle pollution and ground-level ozone are the most widespread health threats. The Environmental Protection Agency calls these two pollutants "criteria" air pollutants, because the EPA regulates these prevailing pollutants by creating human health based and/or environmentally based criteria (science-based guidelines) for setting permissible levels. The set of limits deemed permissible for exposure, based on securing optimal human health, is called the primary standard. The name for another set of limits intended to prevent environmental and property damage, which is mostly used for the other for most common air pollutants, as well as other moderately worrisome air pollutants, is known as the secondary standard.

To remain cognizant of the effects of each of these six common air pollutants, the EPA tracks two kinds of air pollution trends. The first trend involves the air concentration, which is based on actual measurements of pollutant concentrations in the ambient or outside air at selected monitoring sites throughout the country. The second involves the emissions of the air pollutants, which is based on engineering estimates of the total tons of pollutants released into the air each year. Despite the progress made in the last few decades, millions of people continue to live in counties throughout the United States with monitor data showing unhealthy air for one or more of the six common air pollutants. This is alarming for two reasons: the first being that not enough information is in circulation concerning these health hazards, or not enough of it has been made public knowledge. The second concern, which is possibly graver, is that people simply do not concern themselves enough with how potentially devastating these air pollutants can be to themselves, and to the environment.

Air Pollutant Concentrations

Air pollutant concentrations, as measured or as calculated by air pollution dispersion modeling, must often be converted or corrected to be expressed as required by the regulations issued by various governmental agencies. Regulations that define and limit the concentration of pollutants in the ambient air or in gaseous emissions to the ambient air are issued by various national and state (or provincial) environmental protection and occupational health and safety agencies.

Such regulations involve a number of different expressions of concentration. Some express the concentrations as parts per million by volume (ppmv) and some express the concentrations as milligrams per cubic meter (mg/m³), while others require adjusting or correcting the concentrations to reference conditions of moisture content, oxygen content or carbon dioxide content.

The correction of concentrations to some specified reference conditions is most often used in regulations that limit the emissions of particulate matter, NO_x, SO_2 and particulate matter and other gaseous pollutants.

Converting Air Pollutant Concentrations

The conversion equations depend on the temperature at which the conversion is wanted (usually about 20 to 25 °C). At an ambient sea level pressure of one atmosphere, abbreviated atm (101.325 kPa):

$$\text{ppmv} = \text{mg}/\,\text{m}^3 \cdot \frac{(0.08205 \cdot \text{T})}{\text{M}}$$

And for reverse conversion:

$$\text{mg}/\,\text{m}^3 = \text{ppmv} \cdot \frac{\text{M}}{(0.08205 \cdot \text{T})}$$

where,

- $\text{mg}/\,\text{m}^3$ = milligrams of pollutant per cubic meter of air at sea level atmospheric pressure and T.

- ppmv = air pollutant concentration, in parts per million by volume.

- T =ambient temperature in K= 273.15 +°C.

- 0.08205 = universal gas constant in atm-m³/(kmol-k).

- M =molecular weight of air pollutant.

- One atm = absolute pressure of 101.325 kPa.

- mol = gram mole and kmol = 1000 gram moles.

- Air pollution regulations in the United States typically reference their pollutant limits to an ambient temperature of 20 to 25 °C as noted above. In most other nations, the reference ambient temperature for pollutant limits may be 0 °C or other values.

- Although ppmv and mg/m³ have been used for the examples in all of the following sections, concentrations such as ppbv (i.e., parts per billion by volume), volume percent, mole percent and many others may also be used for gaseous pollutants.

- Particulate matter (PM) in the atmospheric air or in any other gas cannot be expressed in terms of ppmv, ppbv, volume percent or mole percent. PM is most usually (but not always) expressed as mg/m³ of air or other gas at a specified temperature and pressure.

- For gases, volume percent = mole percent.

- One volume percent = 10,000 ppmv (i.e. parts per million by volume) with a million being defined as 10^6.

- Care must be taken with the concentrations expressed as parts per billion by volume (ppbv) to differentiate between the British billion which is 10^{12} and the USA billion which is 10^9 (also referred to as the long scale and short scale billion, respectively).

Correcting Concentrations for Altitude

Air pollutant concentrations expressed as mass per unit volume of atmospheric air (e.g., mg/m³, µg/m³) at sea level will decrease with increasing altitude. The concentration decrease is directly proportional to the pressure decrease with increasing altitude. Some governmental regulatory jurisdictions require industrial sources of air pollution to comply with sea level standards corrected for altitude. In other words, industrial air pollution sources located at altitudes well above sea level must comply with significantly more stringent air quality standards than sources located at sea level. For example, New Mexico's Department of the Environment has a regulation with such a requirement.

The derivation of an equation for relating atmospheric pressure to altitude has been published by the Portland State Aerospace Society and it can be rearranged and used as follows:

$$P_h = P_0 \cdot \left(\frac{T_0 - L.h}{T_0} \right)^{\frac{g \cdot M}{R \cdot L}}$$

which, for the parameter values below, then becomes:

$$P_h = P_0 \cdot \left(\frac{T_0 - 6.5h}{T_0} \right)^{5.2558}$$

Given an air pollutant concentration at sea-level atmospheric pressure, the concentration at higher altitudes can be obtained from this equation:

$$C_h = C_0 \cdot \left(\frac{T_0 - 6.5h}{T_0} \right)^{5.2558}$$

where,

- L =atmospheric temperature lapse rate=6.5K/km.

- h = altitude, km.

- g = Earth's surface gravitational acceleration= 9.80665m/s².

- M =Molecular weight of air =28.9644g/mol.

- R =Universal gas constant=8.314472 J/(mol.K).

- P_0 = absolute atmospheric pressure at sea level (in any selected pressure units).

- P_h = absolute atmospheric pressure altitude h (in the same pressure units as P_0).

- C_0 = air pollutant concentration mass/ unit volume at sea level atmospheric preassure P_0 and specified temperature T_0 (commonly 288.15K).

- C_h = concentration, mass / unit volume at altitude h and specified temperature T_0.

As an example, given an air pollutant concentration of 260 mg/m3 at sea level, the equivalent pollutant concentration at an altitude of 2800 meters (2.8 km) is:

$$C_h = 260 \times \left[\left\{ 288 - (6.5)(2.8) \right\} / 288 \right]^{5.2558} = 260 \times 0.71 = 185 \ \text{mg/ m}^3$$

The above equation for the decrease of air pollution concentrations with increasing altitude is applicable only for about the first 10 km of altitude in the troposphere (the lowest atmospheric layer) and is estimated to have a maximum error of approximately three percent. However, 10 km of altitude is sufficient for most purposes involving air pollutant concentrations.

Correcting Concentrations for Reference Conditions

Many environmental protection agencies have issued regulations that limit the concentration of pollutants in air pollution emissions and define the reference conditions applicable to those concentration limits. For example, such a regulation might limit the concentration of nitrogen oxides (NO_x) to 55 ppmv in a dry combustion flue gas (at a specified reference temperature and pressure) corrected to three volume percent of oxygen (O_2) in the dry gas. As another example, a regulation might limit the concentration of total particulate matter to 200 mg/m$_3$ of an emitted gas (at a specified reference temperature and pressure) corrected to a dry basis and further corrected to 12 volume percent carbon dioxide (CO_2) in the dry gas.

Environmental agencies in the USA often use the terms "dscf" or "scfd" to denote a "standard" cubic foot of dry gas. Likewise, they often use the terms "dscm" or "scmd" to denote a "standard" cubic meter of gas. Since there is no universally accepted set of "standard" temperature and pressure, such usage can be and is very confusing. It is strongly recommended that the reference temperature and pressure always be clearly specified when stating gas volumes or gas flow rates.

Correcting to a Dry Basis

If a gaseous emission sample is analyzed and found to contain water vapor and a pollutant concentration of say 40 ppmv, then 40 ppmv should be designated as the "wet basis" pollutant concentration. The following equation can be used to correct the measured "wet basis" concentration to a "dry basis" concentration:

$$C_{dry\,basis} = \frac{C_{wet\,basis}}{1-w}$$

where:

- C = concentration of the air pollutant in the emitted gas.

- w =fraction, by volume, of the emitted gas that is water vapour.

As an example, a wet basis concentration of 40 ppmv in a gas having 10 volume percent water vapor would have an equivalent dry basis concentration of:

$$C_{dry\,basis} = 40 \div (1-0.10) = 44.4\ ppmv$$

Correcting to a Reference Oxygen Content

The following equation can be used to correct a measured pollutant concentration in a dry emitted gas with a measured O_2 content to an equivalent pollutant concentration in a dry emitted gas with a specified reference amount of O_2:

$$C_r = C_m \cdot \frac{(20.9 - \text{reference volume}\,\%\,O_2)}{((20.9 - \text{measured volume}\,\%\,O_2)}$$

where:

- C_r = corrected concentration of a dry gas with a specified reference Volume % O_2.

- C_m = measured concentration in a dry gas having a measured volume % O_2.

For example, when corrected to a dry gas having a specified reference O_2 content of 3 volume %, a measured NO_x concentration of 45 ppmv in a dry gas having a measured 5 volume % O_2 is:

$$C_r = 45 \times (20.9 - 3) \div (20.9 - 5) = 50.7\ ppmv\ of\ NOx$$

The measured gas concentration C_m must first be corrected to a dry basis before using the above equation.

Correcting to a Reference Carbon Dioxide Content

The following equation can be used to correct a measured pollutant concentration in an emitted gas (containing a measured CO_2 content) to an equivalent pollutant concentration in an emitted gas containing a specified reference amount of CO_2:

$$C_r = C_m \cdot \frac{(\text{reference volume}\% \, CO_2)}{(\text{measured volume}\% \, CO_2)}$$

where,

- C_r = corrected concentration of a dry gas with a specified reference Volume % O_2

- C_m = measured concentration in a dry gas having a measured volume % O_2

As an example, when corrected to a dry gas having a specified reference CO_2 content of 12 volume %, a measured particulates concentration of 200 mg/m³ in a dry gas that has a measured 8 volume % CO_2 is:

$$C_r = 200 \times (12 \div 8) = 300 \text{ mg/m}^3$$

The measured gas concentration C_m must first be corrected to a dry basis before using the above equation.

Indoor Air Pollutants

Poor indoor air quality can cause or contribute to the development of infections, lung cancer and chronic lung diseases such as asthma. People who already have lung disease are at greater risk. Following are some indoor air pllutants:

- Asbestos

- Bacteria and viruses

- Building and paint products

- Carbon monoxide

- Carpets

- Cleaning supplies and household chemicals

- Cockroaches

- Dust mites and dust

- Floods and water damage

- Formaldehyde

- Lead

- Mold and dampness

- Nitrogen dioxide

- Pet dander

- Radon

- Residential wood burning

- Secondhand smoke

- Volatile Organic Compounds

Toxic Air Pollutants

Toxic air pollutants, also referred to as hazardous air pollutants, are substances that cause or may cause cancer or other serious health effects, such as reproductive, birth or developmental defects, and neurological, cardiovascular, and respiratory disease. They can be found in gaseous, aerosol, or particulate forms. Some toxic air pollutants (e.g., mercury) are persistent bio accumulative toxics, which means they are stored indefinitely in the body and increase over time. These toxics can deposit onto soils or surface waters, where they are taken up by plants and are ingested by animals, with concentrations increasing as the toxics move up through the food chain to humans. Toxic air pollutants include, among others, formaldehyde; acrolein; benzene; naphthalene; arsenic and metals, such as cadmium, mercury, chromium and lead. Sources of hazardous air pollutants include stationary sources, such as power plants, factories, dry cleaners, and hospitals, as well as mobile sources such as cars, buses, and construction equipment.

What are the Health Effects from Toxic Air Pollutants?

Toxic air pollutants pose different risks to health depending on the specific pollutant, including:

- Cancer, including lung, kidney, bone, stomach.

- Harm to the nervous system and brain.

- Birth defects.

- Irritation to the eyes, nose and throat.

- Coughing and wheezing.

- Impaired lung function.

- Harm to the cardiovascular system.

- Reduced fertility.

How are People Exposed to these Pollutants?

People inhale many of these pollutants in the air where they live. But, since these pollutants also settle into waterways, streams, rivers and lakes, people can drink them in the water or eat them in the fish from these waters. Some hazardous pollutants settle into the dirt that children play in and may put in their mouths.

Where do Toxic Air Pollutants come from?

Major sources of toxic air pollutants outdoors include emissions from coal-fired power plants, industries, and refineries, as well as from cars, trucks and buses.

Indoor air also can contain hazardous air pollutants from sources that include tobacco smoke, building materials like asbestos, and chemicals like solvents.

Criteria Pollutants

The criteria pollutants are carbon monoxide, lead, nitrogen dioxide, ozone, particulate matter, and sulfur dioxide. Criteria pollutants are the only air pollutants with national air quality standards that define allowable concentrations of these substances in ambient air.

Exposure to these substances can cause health effects, environmental effects, and property damage. Health effects include heart or lung disease, respiratory damage, or premature death. Environmental effects include smog, acid rain, radiation, and ozone depletion.

EPA reports that releases of all criteria air pollutants except nitrogen oxides have been in decline since the passage of the 1970 CAA. Overall air quality has improved significantly nationwide since the 1980s. These improvements, however, have not eliminated air quality problems, and major efforts to control pollution sources are still required to ensure everyone breathes air that meets CAA standards.

Carbon Monoxide

Carbon monoxide (CO) is a colorless, odorless, and poisonous gas and one of six criteria pollutants for which EPA has established protective standards.

Lead

Lead is a metal found naturally in the environment as well as in manufactured products. It is one of six criteria pollutants for which EPA has established protective standards.

Ozone

Ozone is a gas that forms in the atmosphere when three atoms of oxygen are combined. It is not emitted directly into the air but is created at ground level by a chemical reaction between oxides of nitrogen, and volatile organic compounds in the presence of sunlight.

Nitrogen Dioxide

Nitrogen dioxide is a brownish, highly reactive gas present in all urban atmospheres. Nitrogen dioxide is a criteria pollutant that can irritate the lungs, cause bronchitis and pneumonia, and lower resistance to respiratory infections.

Particulate Matter

Particulate matter, or PM, is the term for small particles found in the air including dust, dirt, soot, smoke, and liquid droplets. Particles can be suspended in the air for long periods of time.

Sulfur Dioxide

Sulfur dioxide is a colorless, reactive gas produced during burning of sulfur-containing fuels such as coal and oil, during metal smelting, and by other industrial processes.

Short-lived Climate Pollutants

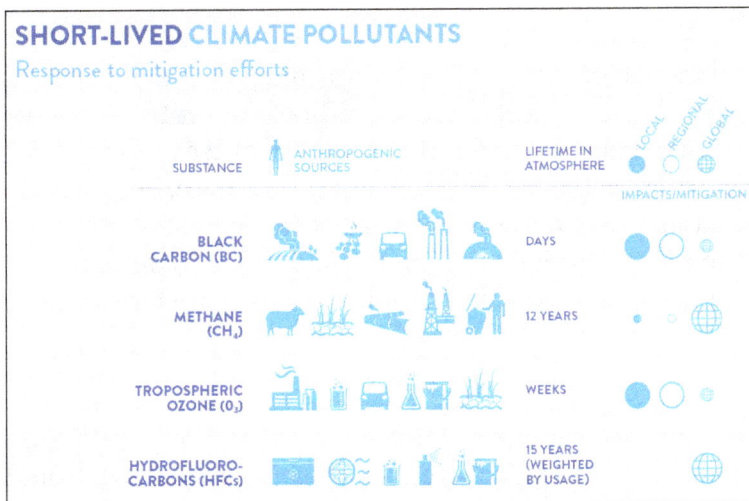

SHORT-LIVED CLIMATE POLLUTANTS
Response to mitigation efforts

SUBSTANCE	ANTHROPOGENIC SOURCES	LIFETIME IN ATMOSPHERE	LOCAL REGIONAL GLOBAL
BLACK CARBON (BC)		DAYS	
METHANE (CH₄)		12 YEARS	
TROPOSPHERIC OZONE (O₃)		WEEKS	
HYDROFLUORO-CARBONS (HFCs)		15 YEARS (WEIGHTED BY USAGE)	

Short-lived climate pollutants are powerful climate forcers that remain in the atmosphere for a much shorter period of time than carbon dioxide (CO_2), yet their potential to warm the atmosphere can be many times greater. Certain short-lived climate pollutants are also dangerous air pollutants that have harmful effects for people, ecosystems and agricultural productivity.

The short-lived climate pollutants black carbon, methane, tropospheric ozone, and hydro fluorocarbons are the most important contributors to the man-made global greenhouse effect after carbon dioxide, responsible for up to 45% of current global warming. If no action to reduce emissions of these pollutants is taken in the coming decades, they are expected to account for as much as half of warming caused by human activity.

Control Measures

According to United Nations Environment (UNEP) and the World Meteorological Organisation (WMO), acting on black carbon and methane in key sectors could reduce projected global warming by 0.5°C by 2050, avoid millions of premature deaths from air pollution annually, prevent millions of tonnes of annual crop losses, and increase energy efficiency, all while providing a number of additional benefits for human wellbeing.

Additionally, fast action under the Montreal Protocol can limit the growth of hydro fluorocarbons (HFCs) and avoid up to 0.5°C of warming by 2100.

Car Pollutants

Car pollutants cause immediate and long-term effects on the environment. Car exhausts emit a wide range of gases and solid matter, causing global warming, acid rain, and harming the environment and human health. Engine noise and fuel spills also cause pollution. Cars, trucks and other forms of transportation are the single largest contributor to air pollution in the United States, but car owners can reduce their vehicle's effects on the environment.

Global Warming

Car pollution is one of the major causes of global warming. Cars and trucks emit carbon dioxide and other greenhouse gases, which contribute one-fifth of the United States' total global warming pollution. Greenhouse gases trap heat in the atmosphere, which causes worldwide temperatures to rise. Without greenhouse gases, the Earth would be covered in ice, but burning excessive amounts of fossil fuels, such as gasoline and diesel, has caused an increase of 0.6 degrees Celsius, or 1 degree F, in global temperatures since pre-industrial times, and this will continue to rise over the coming decades. Warmer global temperatures affect farming, wildlife, sea levels and natural landscapes.

Air, Soil and Water

The effects of car pollution are widespread, affecting air, soil and water quality. Nitrous oxide contributes to the depletion of the ozone layer, which shields the Earth from harmful ultraviolet radiation from the sun. Sulfur dioxide and nitrogen dioxide mix with rainwater to create acid rain, which damages crops, forests and other vegetation and buildings. Oil and fuel spills from cars and trucks seep into the soil near highways, and discarded fuel and particulates from vehicle emissions contaminate lakes, rivers and wetlands.

Human Health

Particulate matter, hydrocarbons, carbon monoxide and other car pollutants harm human health. Diesel engines emit high levels of particulate matter, which are airborne particles of soot and metal. These cause skin and eye irritation and allergies, and very fine particles lodge deep in lungs, where they cause respiratory problems. Hydrocarbons react with nitrogen dioxide and sunlight and form ozone, which is beneficial in the upper atmosphere but harmful at ground level. Ozone inflames lungs, causing chest pains and coughing and making it difficult to breathe. Carbon monoxide, another exhaust gas, is particularly dangerous to infants and people suffering from heart disease because it interferes with the blood's ability to transport oxygen. Other car pollutants that harm human health include sulfur dioxide, benzene and formaldehyde. Noise from cars is also harmful, damaging hearing and causing psychological ill-health.

Reducing Car Pollution

There are several ways that car and truck owners can reduce the effects of car pollutants on the environment. Old and poorly maintained vehicles cause most pollution from cars, but electric, hybrid and other clean, fuel-efficient cars have a reduced impact. When buying a new car, check the fuel economy and environment label. High ratings mean low pollution levels. Maximize fuel economy by removing all unneeded items, such as roof racks, and driving steadily, rather than accelerating quickly and braking hard. Keep your vehicle well-maintained, with regular tune-ups and tire checks, and leave the car at home whenever you can. Walk, bike or use public transportation when possible.

Particulate Matter

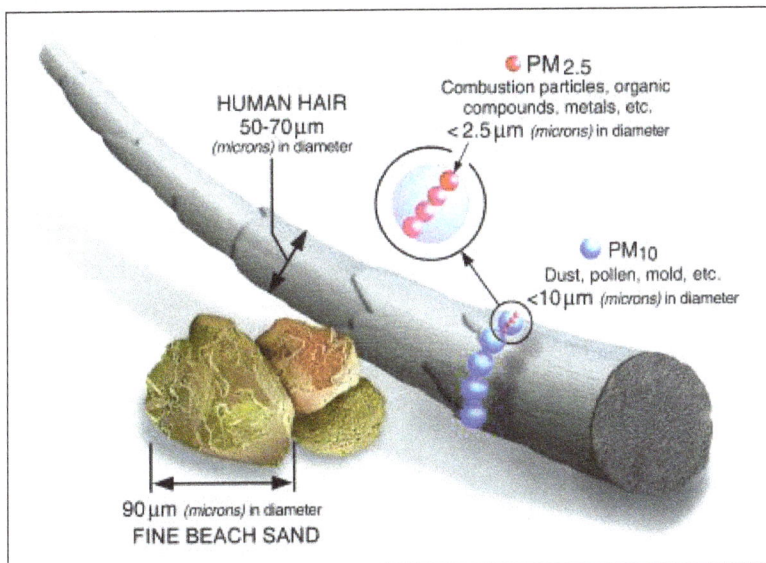

PM stands for particulate matter (also called particle pollution): the term for a mixture of solid particles and liquid droplets found in the air. Some particles, such as dust, dirt, soot, or smoke, are large or dark enough to be seen with the naked eye. Others are so small they can only be detected using an electron microscope.

Particle pollution includes:

- PM_{10}: Inhalable particles, with diameters that are generally 10 micrometers and smaller; and

- $PM_{2.5}$: Fine inhalable particles, with diameters that are generally 2.5 micrometers and smaller.

How small is 2.5 micrometers? Think about a single hair from your head. The average human hair is about 70 micrometers in diameter – making it 30 times larger than the largest fine particle.

Sources of PM

These particles come in many sizes and shapes and can be made up of hundreds of different chemicals.

Some are emitted directly from a source, such as construction sites, unpaved roads, fields, smokestacks or fires.

Most particles form in the atmosphere as a result of complex reactions of chemicals such as sulfur dioxide and nitrogen oxides, which are pollutants emitted from power plants, industries and automobiles.

What are the Harmful Effects of PM?

Particulate matter contains microscopic solids or liquid droplets that are so small that they can be inhaled and cause serious health problems. Some particles less than 10 micrometers in diameter can get deep into your lungs and some may even get into your bloodstream. Of these, particles less than 2.5 micrometers in diameter, also known as fine particles or $PM_{2.5}$, pose the greatest risk to health.

Fine particles are also the main cause of reduced visibility (haze) in parts of the United States, including many of our treasured national parks and wilderness areas.

What is being done to Reduce Particle Pollution?

EPA regulates inhalable particles. Particles of sand and large dust, which are larger than 10 micrometers, are not regulated by EPA.

EPA's national and regional rules to reduce emissions of pollutants that form PM will help state and local governments meet the Agency's national air quality standards.

Nitrogen Oxides

Nitrogen oxides are produced in combustion processes, partly from nitrogen compounds in the fuel, but mostly by direct combination of atmospheric oxygen and nitrogen in flames. Nitrogen oxides are produced naturally by lightning, and also, to a small extent, by microbial processes in soils.

Emission Sources and Trends

Man-made emissions of nitrogen oxides dominate total emissions in Europe, with the UK emitting about 2.2 million tonnes of NO_2 each year. Of this, about one-quarter is from power stations, one-half from motor vehicles, and the rest from other industrial and domestic combustion processes. Unlike emissions of sulphur dioxide, emissions of nitrogen oxides are only falling slowly in the UK, as emission control strategies for stationary and mobile sources are offset by increasing numbers of road vehicles.

- Emissions from electricity generation: NO_x emissions from electricity generation are fairly constant from 1970 onwards until 1990. During the early 1990s the increased use of gas in electricity generation displaced coal and oil. The cleaner fuel and more modern power stations led to a significant reduction in NO_x emissions from the sector until 2000. From 2000, the absolute level of gas used for electricity generation remained fairly constant, and increased demand was met by coal-fired power stations. Since 2006, coal use (and the total amount of fuel used in electricity generation) has substantially decreased.

- NO_x emissions from road transport: The road transport sector has provided a significant contribution to the downward trend in UK emissions. Emissions from road transport currently make the largest contribution to the UK total, accounting for some 33% in 2010. The first petrol cars with three-way catalysts were introduced in 1992, and this resulted in a significant reduction in NO_x emissions. Emission limits for diesel cars and light goods vehicles came into effect in 1993/94. Limits on emissions from heavy goods vehicles (HGVs) first came into effect in 1988 leading to a gradual reduction in emission rates as new HGVs penetrated the fleet. The introduction of these standards has had a substantial impact on NO_x emissions from the road transport sector compared with the 1990s.

Atmospheric Chemistry and Transport

The primary pollutant, directly emitted, is nitric oxide (NO), together with a small proportion of nitrogen dioxide (NO_2). NO is oxidised by ozone in the atmosphere, on a time scale of tens of minutes, to give NO_2. In rural air, away from sources of NO, most of the nitrogen oxides in the atmosphere are in the form of NO_2. NO and NO2 are collectively known as NO_x because they are rapidly inter-converted during the day. NO_2 is split up by UV light to give NO and an O atom, which combines with molecular oxygen (O_2) to give ozone (O_3). Therefore, during the day NO, NO_2 and ozone exist in a quasi-equilibrium which depends on the amount of sunlight. Eventually, NO_2 is oxidised to nitric acid (HNO_3, vapour) which is absorbed directly at the ground, is converted into nitrate-containing particles, or dissolves in cloud droplets. At night, different oxidation processes convert NO_2 to nitrates.

Although nitric acid is rapidly absorbed on contact with surfaces (cloud droplets, soil or vegetation), the other nitrogen oxides are removed only rather slowly, and may travel many hundreds of km before their eventual conversion to nitric acid or nitrates. Consequently, emissions in one country will be deposited in others. The UK exports about three-quarters of its emissions of NO_x.

Measured NO_2 concentrations show the predominance of traffic and urban sources, with the largest concentrations in the large conurbations and adjacent to the motorway network, with annual mean concentrations in excess of 10 ppb in these areas.

Ecosystem Impacts

It is likely that the strongest effect of emissions of nitrogen oxides across the UK is through their contribution to total nitrogen deposition. However, direct effects of gaseous nitrogen oxides, may also be important, especially in areas close to sources (e.g. roadside verges). The critical level for all vegetation types from the effects of NO_x has been set to 30 μg/m³. Experimental evidence suggests that moderate concentrations of NO_x may produce both positive and negative growth responses, with the potential for synergistic interactions with sulphur dioxide (SO_2) being very important. There is substantial evidence to suggest that the effects of NO_2 are much more likely to be negative in the presence of equivalent concentrations of SO_2. At the same time the ratio of SO_2 to NO_2 has decreased greatly in urban areas of the UK over the past 30 years.

One important effect of NO_x may be its influence on insect populations; there is evidence of improved performance of insect pests on plants grown in moderate concentrations of NO_2 and SO_2.

Nitrogen oxides are also one of the precursors for photochemical ozone formation.

Nitrogen Dioxide

Nitrogen Dioxide is a chemical compound with the formula NO_2, but it is usually defined as an indicator for a highly reactive gases group known as oxides of nitrogen or nitrogen oxides (NO_x).

Its smell and color are the only properties of nitrogen dioxide perceptible for humans without any special equipment. NO_2 has a biting and pungent odor and it is easily recognizable with a red-brown color at gas stage (over 21.2 °C) and yellow-brown looking at liquid's (between 21.2 and -11.2 °C).

Nitrogen Dioxide Uses

Nitrogen dioxide is released in wide variety of situations and processes that involve nitrogen. Here are some examples:

- Nitric acid manufacturing.

- Nitrating agent in chemical explosives manufacturing.

- Room temperature sterilization agent.

- Oxidizing rockets fuel.

- Polymerization inhibitor for acrylates.

Sources of Nitrogen Dioxide Pollution

Nitrogen dioxide emissions to the atmosphere are processes that contribute to worsen the air quality, and this is the reason why it is considered criteria pollutant. Nitrogen oxides are produced by human activity 99% of the time, and produced naturally the other 1% during thunderstorms by electric discharge.

Outdoors, cars and combustion engines burning fossil fuels are the number one responsible for nitrogen dioxide emissions. Indoors, NO_2 emissions are mainly produced by sources like cigarettes, butane, kerosene heaters and stoves.

Indirectly, nitrogen monoxide emissions also contribute to the formation of nitrogen dioxide since the first reacts with oxygen or ozone to produce the second.

Nitrogen Dioxide Health and Environmental Effects

As indicator of the NO_x group, nitrogen dioxide is responsible for several health and environmental effects. NO_2 reacts with other gases to create adverse meteorological conditions, such as acid rain or ground-level ozone, known for being a threat to humans and wildlife.

Health Effects on Humans

Nitrogen dioxide, as well as its NO_x siblings, leads to respiratory problems when inhaled since they can penetrate deeply into sensitive lung tissue. Some symptoms are coughing, wheezing or difficulties to breathe.

However, these nitrogen oxides need to react with other compounds like ammonia, volatile organic compounds (VOCs) or common organic chemicals to become extremely harmful, causing then similar health effects than NO_2.

Long-term exposure could carry the development of asthma, emphysema, bronchitis or other respiratory diseases and infections. It can also aggravate cardiovascular problems such as heart diseases. Moreover, in extreme conditions, breathing polluted air with high levels of nitrogen dioxide may even cause premature death.

Sensitive groups such as children, elderly or people with respiratory problems are more affected by the exposure to this pollutant. For these groups, it is recommended controlling NO_x levels and emissions, especially for NO_2 and NO, with devices such as nitrogen dioxide detectors (that can even be portables).

How does Nitrogen Dioxide Pollution affect our Planet?

Nitrogen dioxide's main partner in the NO_x group is nitric oxide or nitrogen monoxide (NO). As already said, both help in the development of environmental effects like smog, acid rain or tropospheric ozone.

Nitrogen dioxide or any others NO_x react with water, oxygen and other chemicals in the atmosphere to form acid rain. Acid rain damages vegetation, buildings, water bodies and all the living beings on these environments.

Despite nitrogen is essential for plants nutrition, high levels of nitrogen dioxide or nitrogen monoxide may damage their lives. Nitrogen oxides in the atmosphere contribute to nutrient pollution in coastal waters and nitrate particles affect the visibility and create hazy air.

How is NO_2 Pollution Controlled?

Over the years, developed countries have reached some agreements to control pollution. Each country or state has its own implementation plan. For example, the Clean Air Act made by the U.S. Environmental Protection Agency (EPA) establishes National Ambient Air Quality Standards (NAAQS) for those criteria pollutants considered most harmful for health and the environment. Another example is the European Union Air Quality Directive by the European Environmental Agency, which also establishes some standards and tips to reduce air pollution, both indoors and outdoors.

Nitrogen dioxide is considered both a primary and secondary criteria pollutant, as it can be extremely dangerous for the environment and the public safety. As mentioned, it acts as the indicator for the nitrogen oxides group, and the maximum permitted or recommended levels for NO_2 are:

- World Health Organization (WHO) guidelines: 200 µg/m³ and 40 µg/m³ for average periods of 1 hour and 1 year, respectively.

- NAAQS: 0,1 ppm of 1-hour daily maximum concentrations, averaged over 3 years, while annual mean cannot exceed 0,053 ppm (100 µg/m³).

EU Air Quality Directive: exactly the same as WHO guidelines, 200 μg/m³ for 1 hour (cannot be exceeded more than 18 times per year) and 40 μg/m³ annually.

Ozone

Ozone (O_3) is present throughout the atmosphere although there are concentration peaks at two levels, the stratosphere (15 - 50 km) and troposphere (0-15 km), with the largest fraction and concentrations being in the stratospheric O_3 layer. Stratospheric O_3 is important as it regulates the transmittance of ultraviolet light to the surface of the earth. Hence reductions in stratospheric O_3 in polar regions, particularly the Antarctic "ozone hole", are of concern regarding the health effects of exposure to increased levels of UV-B.

In contrast, O_3 in the troposphere (ground level) is regionally important as a toxic air pollutant and greenhouse gas. Mixing with stratospheric air provides a natural global average background of around 10-20 parts per billion (ppb), though there is some debate about the concentration. Additional quantities of tropospheric O_3 are produced by photochemical reactions from nitrogen oxides (NO_x) and volatile organic compounds (VOCs), which include various hydrocarbons.

Formation and Sources

Ground-level ozone (O_3) is not emitted directly from anthropogenic sources. It is a "secondary" pollutant formed by a complicated series of chemical reactions in the presence of sunlight. Photochemical reactions of NO_x and VOCs (originating from largely from combustion processes) govern the concentration of ground-level O_3 in the atmosphere. Under typical daytime conditions with a well-mixed atmosphere, three reactions reach equilibrium and no net chemistry occurs.

$$NO + O_3 \;\rightarrow\; NO_2 + O_2$$

$$NO_2 + hn \;\rightarrow\; NO + O$$

$$O + O_2 \;\rightarrow\; O_3 (+M)$$

where,

- hn = sunlight with wavelength 280-430 nm.

- M = any molecule eg N_2 or O_2.

The chemical reactions do not take place instantaneously, but can take hours or days. Ozone levels at a particular location may have arisen from VOC and NO_x emissions many hundreds or even thousands of miles away. Maximum concentrations, therefore, generally occur downwind of the source areas of the precursor pollutant emissions.

Anthropogenic emissions of the ozone precursors (NO_x/VOCs) can also cause large transient increases in ozone concentration, termed episodes or smog. These occur when high concentrations of precursors coincide with weather conditions favourable for ozone production such as when the air is warm and slow moving. These "ozone episodes" provide concentrations of O_3 (>40 ppb) which are toxic both to human health (with a long term threshold objective of 50 ppb daily 8-h mean, WHO) and vegetation.

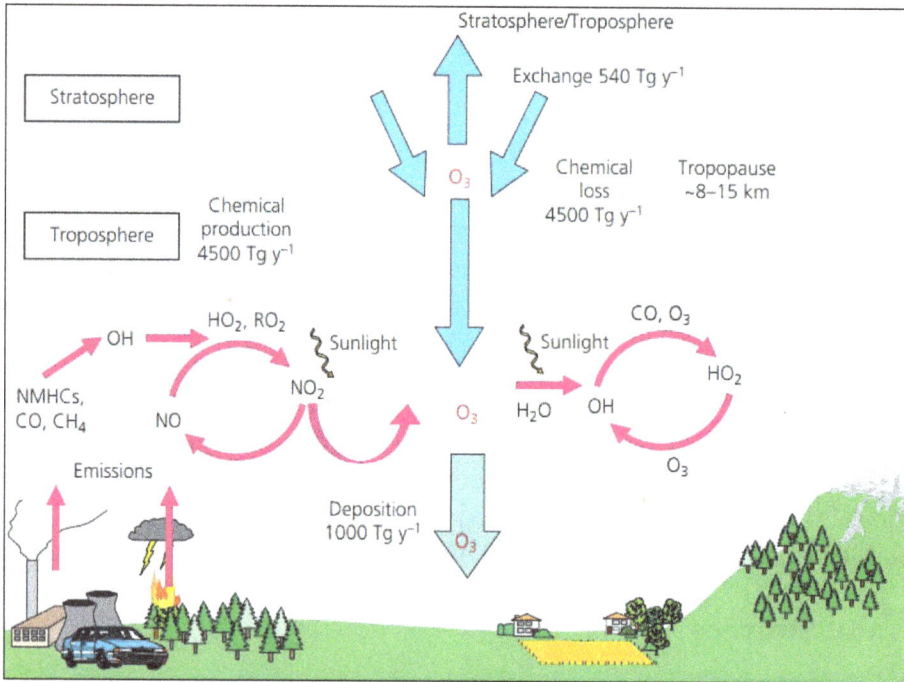

The above image presents a schematic view of the sources and sinks of O_3 in the troposphere. Annual global fluxes of O_3 calculated using a global chemistry–transport model have been included to show the magnitudes of the individual terms. These fluxes include stratosphere to troposphere exchange, chemical production and loss in the troposphere and the deposition flux to terrestrial and marine surfaces.

Prior to the industrial revolution natural sources of NO_x and VOCs would have generated O_3 in the troposphere, adding to that transported from the stratosphere. However, the large amounts of NO_x and VOCs released by human activities, such as the combustion of fossil fuels, has led to a large increase in the northern hemisphere background concentration. Evaluation of historical O_3 measurements indicate that since the 1950s, the background ozone concentration has roughly doubled, although there has been some slowing down of this trend in the last decade.

NO_x		VOC	
Natural	Anthropogenic	Natural	Anthropogenic
Soils, natural fires.	Transport (road, sea and rail), power stations, other industry and combustion processes.	Vegetation, natural fires.	Transport, combustion processes, solvents, oil production.

Concentrations

Policy actions to date across Europe have reduced the emissions of ozone precursors NO_x and VOCs. These emission controls have reduced peak ozone concentrations by typically 30 ppb in the UK, but over the last 20 years mean concentrations have been increasing in urban areas due to reductions in the local depletion of O_3 by NO. Background concentrations have increased in rural areas due to increases in the hemispheric background O_3 concentrations which have increased by approximately 0.2 ppb per year, or by about 5 ppb. The cause of the increases in background O_3 is increases in precursor emissions throughout the northern hemisphere, including shipping, aircraft, vehicle, and industrial emissions in developing economies.

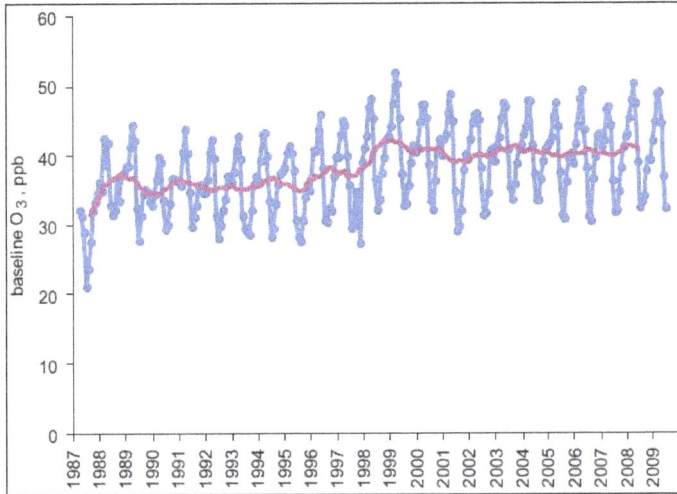

Trends in monthly mean baseline ozone levels at Mace Head on the west coast of Ireland between 1986 and 2006, including a running 12-month mean (pink line), showing the ~5 ppb increase in ozone mixing ratios over 20 years.

Ozone concentrations are highly variable, spatially and temporally. Figure shows the high background concentration in March to May over the northern half of Scotland, upland Wales and northern England, and parts of southern and eastern England. In the summer months of May-July high concentrations cover parts of south-east England and some upland areas of Wales and northern Scotland. The summer time spatial pattern extends across Western Europe and is caused by regions where ozone production occurs more frequently in a combination of precursor emissions (NO_x and VOCs), high solar radiation and temperatures. During the winter the highest concentrations occurring in the NW and the lowest in the SE as a consequence of ozone destruction in the NO polluted atmosphere of the dense urban/industrial regions of the southern UK and continental Western Europe.

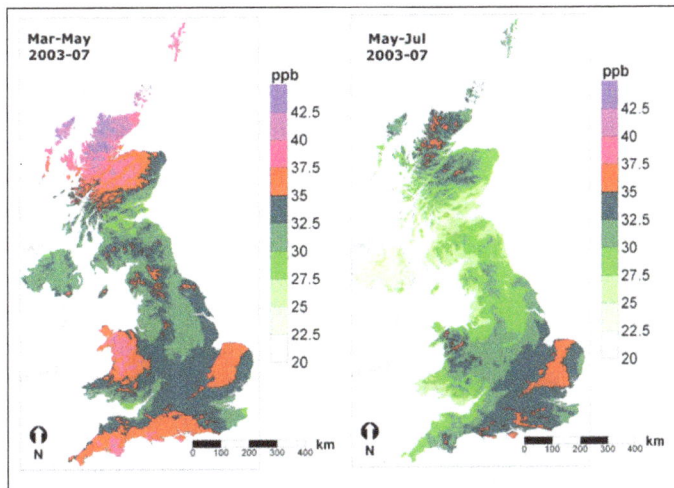

Distribution of mean ozone concentrations over the periods March May and May-July, using data for the five year period 2003-07.

Effects

A large body of evidence has shown that ambient O_3 causes damage to O_3-sensitive vegetation. O_3 enters leaves via the stomatal pores on the leaf surface. Once inside the leaf, a series of chemical reactions occur leading to damage to cell membranes and other negative impacts on plant metabolism, including photosynthesis. These effects can be in response to short-term episodes or cumulative during the growing season, and can lead to:

- Visible leaf damage and premature aging of leaves.

- Reductions in above- and below-ground growth and biomass.

- Changes in the ratio between shoot and root biomass (including carbon allocation).

- Reductions in flower number, flower biomass and seed production.

- Reductions in crop yield quantity and quality, including cereal grains, potato tubers and tomato fruit.

- Changes in forage quantity and quality for pasture.

- Altered tolerance to abiotic stresses such as drought and frost and biotic stresses such as pest attacks and diseases.

Critical Levels

Ozone pollution effects on vegetation are described by cumulative metrics that are based on either atmospheric concentration of O_3 over a threshold concentration (AOT40) or modelled uptake of O_3 through the stomatal pores on the leaf surface (Phytotoxic Ozone Dose above a threshold of Y, POD_Y). A recent analysis of field evidence for O_3 effects confirmed that maps generated using an O_3 flux metric more closely matched locations of damage than those based on O_3 concentration. New O_3 flux-based critical levels have been derived for crops, tree species and for grassland species.

Other Impacts

The role of ozone as a contributor to the direct radiative forcing of global climate has grown in importance. In addition, the recognition that the effects of O_3 on carbon sequestration through its effects on primary productivity of vegetation is an additional reason for interest in ground level ozone.

Ground-level Ozone

Ground-level ozone is a colorless and highly irritating gas that forms just above the earth's surface. It is called a "secondary" pollutant because it is produced when two primary pollutants react in sunlight and stagnant air. These two primary pollutants are nitrogen oxides (NO_x) and volatile organic compounds (VOCs).

NO_x and VOCs come from natural sources as well as human activities. About 95 per cent of NO_x from human activity come from the burning of coal, gasoline and oil in motor vehicles, homes, industries and power plants. VOCs from human activity come mainly from gasoline combustion and marketing, upstream oil and gas production, residential wood combustion, and from the evaporation of liquid fuels and solvents. Significant quantities of VOCs also originate from natural (biogenic) sources such as coniferous forests.

Ozone is known to have significant effects on human health. Exposure to ozone has been linked to pre-mature mortality and a range of morbidity health end-points such as hospital admissions and asthma symptom days. In addition to its effects on human health, ozone can significantly impact vegetation and decrease the productivity of some crops. It can also injure flowers and shrubs and may contribute to forest decline Ozone can also damage synthetic materials, cause cracks in rubber, accelerate fading of dyes, and speed deterioration of some paints and coatings. As well, it damages cotton, acetate, nylon, polyester and other textiles.

Sulfur Dioxide

Sulfur dioxide is the greatest concern for the larger group of gaseous sulfur oxides (SO_x). All standards and control measures for SO_2 include all sulfur oxides, as sulfur dioxide is used as the indicator due to it has a major presence in the air.

Its chemical formula is SO_2, it is invisible and has a nasty, pungent, irritating and sharp smell. Sulfur dioxide easily reacts with other substances to form harmful compounds, such as sulfuric acid, sulfurous acid and sulfate particles.

Sulfur dioxide is used in a big variety of situations, for example:

- To produce acid sulfuric, through a method called contact process.

- To preserve dried apricots, figs or other fruits.

- Used as an antibiotic and antioxidant in winemaking.

- To decolorize substances like swimming pool water, where the blue color of chlorine is removed.

Despite it is a toxic gas, sulfur dioxide is used in many more situations and it has a huge presence in our daily lives.

Sulfur Dioxide Sources

About 99% of sulfur dioxide emissions are produced by industrial activities, such as generation of electricity from coal, oil or gas. It is also produced in the burning of fossil fuels on industrial facilities, by the extraction of metal from ore or by vehicles such as cars, ships or locomotives that burn fuel.

Naturally, SO_2 is formed by volcanic eruptions in active volcanoes. However, it is interesting to know there is a high presence of SO_2 emissions on Venus, Mars and Jupiter, being one of the most significant gases in those atmospheres.

What are the Health and Environmental Effects of Sulfur Dioxide Pollution?

Sulfur dioxide, as well as the others sulfur oxides, have huge impact on the environment and dangerously affect humans, both quite similarly to what other criteria pollutants do.

How SO_2 can Affect Humans Health?

Sulfur dioxide is such a toxic gas that you can feel the firsts symptoms just 10 to 15 minutes after breathing it. Short-term exposure causes problems to the respiratory system, such as breathing difficulties, nose and throat irritating, coughing, wheezing and shortness of breath.

Sensitive groups such as elderly, children or asthmatics will notice strongest symptoms and effects. They are also more susceptible to develop diseases, in case they do not have them yet.

High concentrations of SO_2 in the atmosphere commonly create other SO_x, which react at the same time with other compounds to form small air pollution particles (PM). As a result, long-term exposure can seriously damage your lungs since particulate matter can penetrate deeply into our organisms.

Since one of its uses is to preserve food, bad production process could provoke poisoning due to sulfur dioxide ingestion. Be careful when you buy it.

What is Sulfur Dioxide Pollution Impact on Environment?

This series of events and reactions from sulfur dioxide and other sulfur oxides to create particulate matter may drive to reduce the visibility in open spaces and produce haze. The deposition of these particles may damage stones, buildings, statues and monuments.

High concentrations of SO_x can be harmful for vegetation foliage and growth, and can contribute to acid rain formation, which causes several issues on sensitive ecosystems.

What is being done to Reduce Sulfur Dioxide Pollution?

Over the years, developed countries have reached some agreements to control pollution. Each country or state has its own implementation plan. For example, the Clean Air Act made by the U.S. Environmental Protection Agency (EPA) establishes National Ambient Air Quality Standards (NAAQS) for those criteria pollutants considered most harmful for health and the environment. Another example is the European Union Air Quality Directive by the European Environmental Agency, which also establishes some standards and tips to reduce air pollution, both indoors and outdoors.

Sulfur dioxide, thanks to its high toxicity and danger for humans and environment, is considered both a primary and secondary criteria pollutant. Maximum permitted or recommended levels for SO_2 are:

- NAAQS: 75 ppb per hour and 0.5 ppm (1,300 µg/m³) every 3 hours (which cannot be exceeded more than once per year).

- EU Air Quality Directive: 250 µg/m³ each hour and 125 µg/m³ daily, what cannot be exceeded more than 24 and 3 times per year respectively.

Carbon Monoxide

Carbon monoxide (CO)—a colorless, odorless, tasteless, and toxic air pollutant—is produced in the incomplete combustion of carbon-containing fuels, such as gasoline, natural gas, oil, coal, and wood. The largest anthropogenic source of CO in the United States is vehicle emissions. Breathing the high concentrations of CO typical of a polluted environment leads to reduced oxygen (O_2) transport by hemoglobin and has health effects that include headaches, increased risk of chest pain for persons with heart disease, and impaired reaction timing. In the 1960s, vehicle emissions led to increased and unhealthful ambient CO concentrations in many U.S. cities. With the introduction of emissions controls, particularly automotive catalysts, estimated CO emissions from all sources decreased by 21% from 1980 to 1999. Average ambient concentrations decreased by about 57% over the same period.

The locations that continue to have high concentrations of CO tend to have topographical or meteorological characteristics that exacerbate pollution; for example, strong temperature inversions or the existence of nearby hills that inhibit wind flow may limit pollutant dispersion. Because of the limited dispersion, many of those areas also have unhealthful concentrations of summer ozone (O_3) and year-round particulate matter (PM).[1] Low temperatures also contribute to high CO concentrations. Engines and vehicle emissions-control equipment operate less efficiently when cold: Air-to-fuel ratios are lower, combustion is less complete, and catalysts take longer to become fully operational. The result is that products of incomplete combustion, including CO, are formed in higher concentrations. Sometimes, topography, meteorology, and emissions combine to cause high concentrations of CO.

Health Effects of CO

Clinical and Epidemiological Studies of CO Effects

CO affects human health by impairing the ability of the blood to bring O_2 to body tissues. When CO is inhaled, it rapidly crosses the alveolar epithelium to reach the blood, where it binds to hemoglobin to form carboxyhemoglobin (COHb), a useful marker for predicting the health effects of CO. Because CO has an affinity for hemoglobin more than 200 times greater than does O_2, the presence of CO in the lung will displace O_2 from the hemoglobin. In other words, when CO is present in the lungs, the hemoglobin will be unable to reach 100% O_2 saturation. In addition, the presence of COHb increases hemoglobin's affinity for O_2, thereby inhibiting release of O_2 from the hemoglobin to body tissues. The effect of COHb is illustrated by a leftward shift of the O_2-hemoglobin dissociation curve. As shown in figure, once COHb forms, the hemoglobin is unable to reach 0% O_2 saturation. This second effect continues until the COHb dissociates, typically several hours after CO exposure. CO not only decreases the O_2-carrying capacity of the blood, but also decreases the ability of the tissues to extract O_2 from the blood during circulation. CO has also been shown to bind to myoglobin and may affect O_2 transport to muscle.

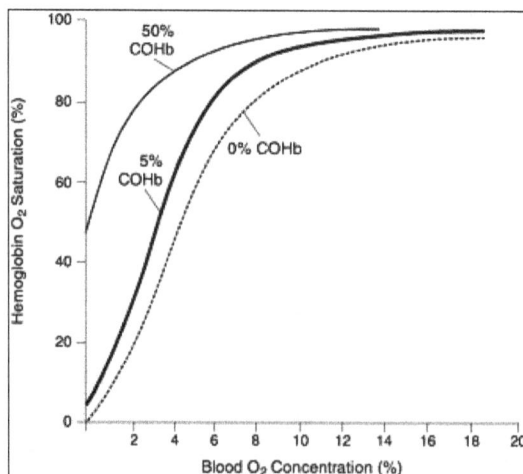

Diagram of hemoglobin response to the presence of COHb. The concentration of O_2 in the environment surrounding the hemoglobin is shown on the x-axis. The O_2 saturation, or how much of the hemoglobin's capacity for storing O_2 is used, is shown on the y-axis. At higher O_2 concentrations, as are found in the lungs, the hemoglobin can be more O_2 saturated. Likewise, at lower O_2 concentrations, as are found in other parts of the body, O_2 will dissociate from the hemoglobin to achieve O_2 saturations as indicated by the curve. The presence of COHb shifts this curve to the left. For a given O_2 concentration, the hemoglobin will require a higher O_2 saturation and allow less O_2 to be released to body's tissues.

COHb levels in healthy individuals not recently exposed to high concentrations of ambient CO are 0.3 to 0.7%. Exposure to high concentrations of ambient CO can result in concentrations of COHb of 2% or higher if the exposure lasts long enough (hours). For those who smoke, cigarette-smoking is typically the most significant source of personal CO exposure. COHb concentrations, which are generally less than 1% in nonsmokers, average about 5% in smokers and are up to 10% or even higher in some very heavy smokers.

The CO health standards set by EPA are intended to keep COHb concentrations for nonsmokers below 2% in order to protect the most susceptible members of the population.

The acute affects of CO poisoning are well understood. Generally, in otherwise healthy people, headache develops when COHb concentrations reach 10%; tinnitus (ringing in the ear) and light-headedness at 20%; nausea, vomiting, and weakness at 20–30%; clouding of consciousness and coma at around 35%; and death at around 50%. However, the outcomes of long-term, low-concentration CO exposures are less well understood. Because of the critical nature of blood flow and O_2 delivery to the heart and brain, these organ systems, as well as the lungs (the first organ to come into contact with the pollutant), have received the most attention.

In patients with known coronary artery disease, COHb concentrations as low as 3% exacerbate the development of exercise-induced chest pain. Concentrations as low as 6% are associated with an increase in the number and frequency of premature ventricular contractions during exercise in patients with severe heart disease. Large environmental-exposure cohort studies have confirmed that daily increases in ambient CO concentrations are associated with statistically significant increases in the numbers of hospital admissions for heart disease and congestive heart failure and with increases in deaths from cardiopulmonary illnesses.

Neuropsychiatric (neurological and psychiatric) disorders and cognitive impairments due to long-term, low-concentration CO exposures have been hypothesized in part on the basis of extrapolation from the known acute effects of high-dose CO poisoning and the concomitant subacute and delayed neuropsychological sequelae. In clinical experiments on healthy volunteers, controlled CO exposure was associated with subtle alterations in visual perception when COHb concentrations were above 5%. However the significance of this finding remains unknown. Similar studies have shown measurable but small effects on auditory perception, driving performance, and vigilance.

The role of CO in pulmonary disease is unclear. In the Seattle area, a single-pollutant model showed a 6% increase in the rate of hospital admissions for asthma with each 0.9-ppm increase in CO, but that was concomitant with increases in other air pollutants. In Minneapolis and Toronto, CO concentrations showed only weak and inconsistent associations with total admissions for respiratory diseases.

A fetus is more susceptible to CO than an adult; the O_2-hemoglobin dissociation curve is to the left of that in the adult. It is shifted even further to the left by CO exposure. Also, because the half-life of fetal COHb is longer than that of adults, it may take up to five times longer to reduce the concentrations to normal. Other studies have shown that exposure to high concentrations of CO during the last trimester of pregnancy may increase the risk of low birth weights and that exposures to CO and PM during pregnancy may trigger preterm births.

Public-health laws are designed to protect the most susceptible people in the population. People with coronary artery disease or other cardiopulmonary diseases, fetuses, infants, and athletes who exercise heavily in high-CO atmospheres are particularly susceptible to experiencing adverse health effects from CO. The evidence summarized above, and described more fully by EPA, indicates that attainment of the ambient-CO standards can decrease morbidity and mortality from atherosclerotic heart disease. Although less conclusive, there is evidence that attainment of the CO standards will also decrease morbidity from pulmonary disease, neurological disease, fetal loss, and childhood developmental abnormalities. These health benefits translate into economic savings associated with avoided health care and avoided work-time losses as well as intangible savings in life quality.

CO Exposure

Motor-vehicle emissions are the primary source of CO in outdoor air in populated areas and are associated with the highest outdoor CO exposure in nonsmokers. Outdoor concentrations of CO tend to be higher in urban areas and to increase with the density of vehicles and miles driven. Measurements of ambient CO typically exhibit a bimodal diurnal pattern, with the highest concentrations generally occurring on weekdays during the commuting hours of 7:00–9:00 a.m. and 4:00–6:00 p.m.[2] CO also accumulates in the rider compartments of motor vehicles. Studies have shown that when the concentration near roadways averages 3–4 ppm, the average concentration in the cab is typically 5 ppm.

Most people spend a majority of their time indoors; this is particularly true in Fairbanks and other cold climates during the winter, when ambient CO concentrations tend to be highest. That leads to the question of the relationship between indoor and outdoor concentrations. Air pollution in buildings can come from indoor sources and from air exchange with outdoor ambient pollution. Air exchange may be active, as in the case of a mechanical ventilation system, or passive, as in the case of infiltration associated with temperature or pressure differences between the outside and the interior of a building. CO penetrates freely with infiltration air from the outside and is not removed by building materials or ventilation systems. Furthermore, there are no effective indoor chemical or physical processes for lowering CO on the time scales of interest for exposure and toxic effects.

The relationship between indoor and outdoor CO concentrations can be evaluated with a simple differential mass-balance model that has the following steady-state solution when we combine active ventilation and passive infiltration into a single air-exchange term:

$$C_i = \frac{paC_0}{a+k} + \frac{S}{(a+k)V},$$

where,

- C_i =indoor concentration, $\mu g/m^3$;
- C_0 =outdoor concentration, $\mu g/m^3$;
- p =penetration coefficient, 0–1;
- a =air exchange rate, h^{-1};
- k =decay rate, h^{-1};
- S =mass flux of the indoor source, $\mu g/h$; and
- V =building volume, m^3.

Although most areas show a bimodal diurnal pattern with respect to ambient CO concentrations, Fairbanks, Alaska, the focus of this interim report, typically shows a continuous increase throughout the day, with 1-h average CO concentrations peaking at 5:00–6:00 p.m.

For CO, the relationship is simpler because the penetration coefficient (p) is unity and the decay rate (k) is effectively zero. Therefore, the solution is,

$$C_i = C_0 + \frac{S}{aV}.$$

It is clear that in the absence of indoor sources (S), the steady-state indoor concentration of CO will equal the outdoor concentration. When a source of CO is present indoors (for example, from a faulty furnace, an underground parking garage, a kerosene heater, or a smoker), the indoor source adds to the background concentration from the outdoor air. Therefore, buildings do not provide protection from high outdoor concentrations of CO. The idea that buildings provide protection from high outdoor CO concentration is a common misconception.

Related Pollutants

The incomplete combustion of fossil fuels, which is responsible for CO emissions, also causes emissions of fine particles ($PM_{2.5}$) and toxic organic air contaminants. Epidemiological studies have linked exposure to $PM_{2.5}$ with various adverse health effects, including premature mortality, exacerbation of asthma and other respiratory tract diseases, and decreased lung function. Because of these adverse health effects, EPA issued NAAQS regulating ambient concentrations of $PM_{2.5}$. In addition to the six criteria air pollutants previously regulated, the 1990 amendments of the Clean Air Act (CAAA90) designated 189 toxic air contaminants. Incomplete combustion in mobile sources is estimated to contribute a substantial fraction to the emissions of several toxic air pollutants, including benzene, 1,3-butadiene, and aldehydes. Each of these toxic air pollutants poses some carcinogenic risk. In addition, chronic exposure to benzene is associated with blood disorders; chronic exposure to 1,3-butadiene is associated with cardiovascular disease; and chronic exposure to aldehydes is associated with respiratory problems and eye, nose, and throat irritation (EPA 1994). Many emissions-control strategies for CO will also cause reductions in these copollutants and their associated adverse health effects.

In this regard, CO is different from O_3, which is highly reactive and therefore rapidly destroyed when infiltrating inside from outdoors.

CO may be a good indicator gas for other pollutants that are emitted at the same time but are not widely measured. The concentrations and spatial distributions of the copollutant species are generally not as well known as for CO. In particular, little data are available about exposure to air toxics present in the ambient environment. CO could be especially useful as an indicator of mobile-source emissions of $PM_{2.5}$ and air toxics, which some studies have shown to be strongly correlated with CO. In one study of emissions from in-use vehicles sampled in Denver, San Antonio, and the Los Angeles area, strong correlations were found between CO and particle emissions ($R^2=0.65$) and between particle and total hydrocarbon (HC) emissions ($R^2=0.78$). The same study demonstrates that emissions of the pollutant species increase with vehicle age and during cold starts. For individual vehicles, however, the correlation among the pollutants is weaker, reflecting the complex mechanisms of formation of the related combustion products.

Volatile Organic Compounds

Volatile organic compounds (VOCs) are emitted as gases from certain solids or liquids. VOCs include a variety of chemicals, some of which may have short- and long-term adverse health effects. Concentrations of many VOCs are consistently higher indoors (up to ten times higher) than outdoors. VOCs are emitted by a wide array of products numbering in the thousands.

Organic chemicals are widely used as ingredients in household products. Paints, varnishes and wax all contain organic solvents, as do many cleaning, disinfecting, cosmetic, degreasing and hobby products. Fuels are made up of organic chemicals. All of these products can release organic compounds while you are using them, and, to some degree, when they are stored.

Sources of VOCs

Household products, including:

- Paints, paint strippers and other solvents,

- Wood preservatives,

- Aerosol sprays,

- Cleansers and disinfectants,

- Moth repellents and air fresheners,

- Stored fuels and automotive products,

- Hobby supplies,

- Dry-cleaned clothing,

- Pesticide,

Other products, including:

- Building materials and furnishings,

- Office equipment such as copiers and printers, correction fluids and carbonless copy paper,

- Graphics and craft materials including glues and adhesives, permanent markers and photographic solutions.

Health Effects

Health effects may include:

- Eye, nose and throat irritation,

- Headaches, loss of coordination and nausea,

- Damage to liver, kidney and central nervous system,

- Some organics can cause cancer in animals, some are suspected or known to cause cancer in humans.

Key signs or symptoms associated with exposure to VOCs include:

- Conjunctival irritation,

- Nose and throat discomfort,

- Headache,

- Allergic skin reaction,

- Dyspnea,

- Declines in serum cholinesterase levels,

- Nausea,

- Emesis,

- Epistaxis,

- Fatigue,

- Dizziness.

The ability of organic chemicals to cause health effects varies greatly from those that are highly toxic, to those with no known health effect.

As with other pollutants, the extent and nature of the health effect will depend on many factors including level of exposure and length of time exposed. Among the immediate symptoms that some people have experienced soon after exposure to some organics include:

- Eye and respiratory tract irritation,

- Headaches,

- Dizziness,

- Visual disorders and memory impairment.

At present, not much is known about what health effects occur from the levels of organics usually found in homes.

Levels in Homes

Studies have found that levels of several organics average 2 to 5 times higher indoors than out-doors. During and for several hours immediately after certain activities, such as paint stripping, levels may be 1,000 times background outdoor levels.

Steps to Reduce Exposure

- Increase ventilation when using products that emit VOCs.

- Meet or exceed any label precautions.

- Do not store opened containers of unused paints and similar materials within the school.

- Formaldehyde, one of the best known VOCs, is one of the few indoor air pollutants that can be readily measured.

    ○ Identify, and if possible, remove the source.

    ○ If not possible to remove, reduce exposure by using a sealant on all exposed surfaces of paneling and other furnishings.

- Use integrated pest management techniques to reduce the need for pesticides.

- Use household products according to manufacturer's directions.

- Make sure you provide plenty of fresh air when using these products.

- Throw away unused or little-used containers safely; buy in quantities that you will use soon.

- Keep out of reach of children and pets.

- Never mix household care products unless directed on the label.

Follow Label Instructions Carefully

Potentially hazardous products often have warnings aimed at reducing exposure of the user. For example, if a label says to use the product in a well-ventilated area, go outdoors or in areas equipped with an exhaust fan to use it. Otherwise, open up windows to provide the maximum amount of outdoor air possible.

Throw away Partially Full Containers of old or Unneeded Chemicals Safely

Because gases can leak even from closed containers, this single step could help lower concentrations of organic chemicals in your home. (Be sure that materials you decide to keep are stored not only in a well-ventilated area but are also safely out of reach of children.) Do not simply toss these unwanted products in the garbage can. Find out if your local government or any organization in your community sponsors special days for the collection of toxic household wastes. If such days are available, use them to dispose of the unwanted containers safely. If no such collection days are available, think about organizing one.

Buy Limited Quantities

If you use products only occasionally or seasonally, such as paints, paint strippers and kerosene for space heaters or gasoline for lawn mowers, buy only as much as you will use right away.

Keep Exposure to Emissions from Products Containing Methylene Chloride to a Minimum

Consumer products that contain methylene chloride include paint strippers, adhesive removers and aerosol spray paints. Methylene chloride is known to cause cancer in animals. Also, methylene chloride is converted to carbon monoxide in the body and can cause symptoms associated with exposure to carbon monoxide. Carefully read the labels containing health hazard information and cautions on the proper use of these products. Use products that contain methylene chloride out-doors when possible; use indoors only if the area is well ventilated.

Keep Exposure to Benzene to a Minimum

Benzene is a known human carcinogen. The main indoor sources of this chemical are:

- Environmental tobacco smoke,
- Stored Fuels,
- Paint supplies,
- Automobile emissions in attached garages.

Actions that will reduce benzene exposure include:

- Eliminating smoking within the home,
- Providing for maximum ventilation during painting,
- Discarding paint supplies and special fuels that will not be used immediately.

Keep Exposure to Perchloroethylene Emissions from Newly Dry-cleaned Materials to a Minimum

Perchloroethylene is the chemical most widely used in dry cleaning. In laboratory studies, it has been shown to cause cancer in animals. Recent studies indicate that people breathe low levels of this chemical both in homes where dry-cleaned goods are stored and as they wear dry-cleaned clothing. Dry cleaners recapture the perchloroethylene during the dry-cleaning process so they can save money by re-using it, and they remove more of the chemical during the pressing and finishing processes. Some dry cleaners, however, do not remove as much perchloroethylene as possible all of the time.

Taking steps to minimize your exposure to this chemical is prudent:

- If dry-cleaned goods have a strong chemical odor when you pick them up, do not accept them until they have been properly dried.

- If goods with a chemical odor are returned to you on subsequent visits, try a different dry cleaner.

Carbon Disulphide

Pure Carbon disulfide is a colourless liquid which has a pleasant smell. However, it usually contains impurities which give it a yellow colour and a less pleasant smell. Carbon disulfide liquid evaporates easily. Its gas is explosive at high concentrations. Carbon disulfide is one of a group of chemicals known as the volatile organic compounds (VOCs).

Carbon disulfide is mainly used in the production of synthetic fibres such as cellophane, rayon and viscose. It is also used in the manufacture of rubbers, agricultural and other chemicals and it is used as a solvent (particularly in the petroleum industry). Naturally occurring Carbon disulfide plays an important role in the global cycle of sulfur exchange between land, oceans and the atmosphere.

The most significant releases of Carbon disulfide are from industry producing or using it. Small amounts are released naturally from volcanoes and marshland.

Due to the ease with which it evaporates, releases of Carbon disulfide tend to end up in the atmopshere. It breaks down quickly and so is not transported or accumulated. Only very high concentrations might harm the environment in the vicinity of its release. As a VOC, Carbon disulfide can be involved in the formation of ground level ozone which can cause damage to crops and materials. It is not considered likely that Carbon disulfide pollution has any effects on the global environment.

How might Exposure to it Affect Human Health?

Carbon disulphide exposure occurs mainly in the occupational setting, and can enter the body either by inhalation of air containing carbon disulphide vapours, by ingestion of contaminated water or food or by dermal contact with soil, water or substances containing carbon disulphide. Inhalation of air containing carbon disulphide can lead to a number of adverse health effects such as chest pains, muscle pain, loss of feeling in hands and feet, loss of memory, weight loss, and liver and kidney damage. Exposure may also affect fertility in both men and women. Inhalation of high levels of carbon disulphide and exposure over long periods of time can result in damage to the heart and central nervous system and in extreme cases death. Ingestion of carbon disulphide may cause irritation of the digestive tract and effects similar to those caused by inhalation. Dermal contact with carbon disulphide can lead to skin irritation and sensitisation. Exposure to high concentrations can cause skin burns. If adsorbed through the skin similar symptoms to those for inhalation may occur. The International Agency for Research on Cancer has not classified carbon disulphide in terms of its carcinogenicity to humans. However, exposure to carbon disulphide at normal background levels is unlikely to have any adverse effect on human health.

Mercury

When the energy sector releases mercury (Hg) as a pollutant, it creates environmental problems. Both humans and natural sources release mercury; burning coal specifically releases quite a bit of mercury. Human activity contribute anywhere from 50-90% of the mercury present in the environment.

The estimates are not more precise because it is difficult to figure out how much re-emitted mercury was initially emitted by humans. Re-emission occurs when previously stored mercury is re-introduced into the environment by forest fires or other means, and requires complex modelling techniques to determine how much can be traced back to human emissions.

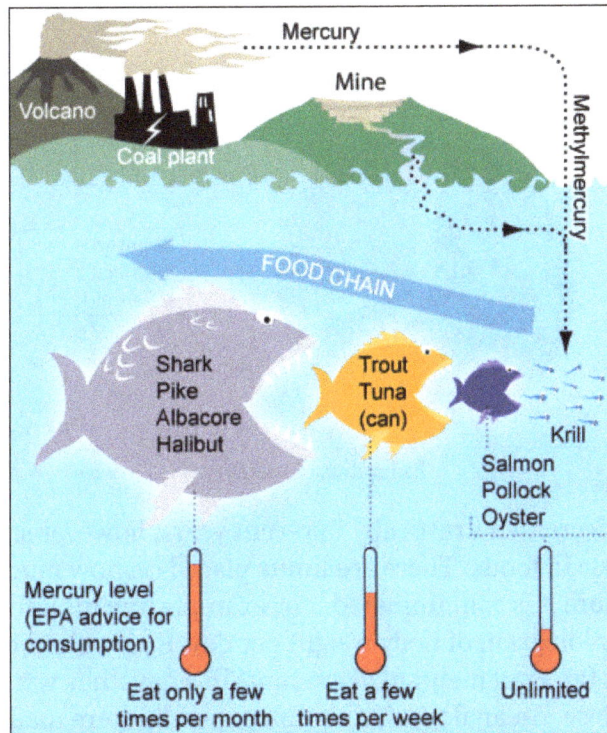

Mercury often finds its way into water bodies and into small fish. Through the process of biomagnification, the concentrations of mercury rise to very toxic levels. This makes shark and pike more hazardous to eat than canned tuna which is still more hazardous than salmon.

Emissions

Mercury levels in the upper layers of oceans are much higher than they were in the past, at approximately double what they were in the pre-industrial era. This is caused primarily by human emissions, which rise into the atmosphere and fall out into soils and bodies of water.

Mercury is released naturally from rocks, soil, volcanoes, and by vaporization from the ocean. This contributes to about 10% of the global input of mercury into the atmosphere.

In addition to coal burning, humans emit mercury with mining and smelting, cement production, oil refining, gold mining, and wastes from consumer products. Asia contributes to about half of the total human input of mercury because of the extensive coal burning for electricity.

The largest emission source contributor for mercury is re-emission. This means that mercury that was previously deposited from the air onto soils, surface waters, and vegetation from past emissions can be emitted back into the air via forest fires or biomass burning. Mercury may be deposited and re-emitted many times as it cycles throughout the environment, potentially cycling indefinitely.

Re-emitted mercury should not be considered natural in origin, even though it may have originally come from a natural source. As humans have introduced more mercury into the environment, re-emissions of mercury have also increased due to environmental burdens caused by the higher levels of input. Continuing to add to the mercury levels will result in ever increasing levels of re-emission, and even when human emissions are cut out completely, this mercury could remain in the system for a long time. This is why it is important that continued efforts are made to reduce mercury pollution.

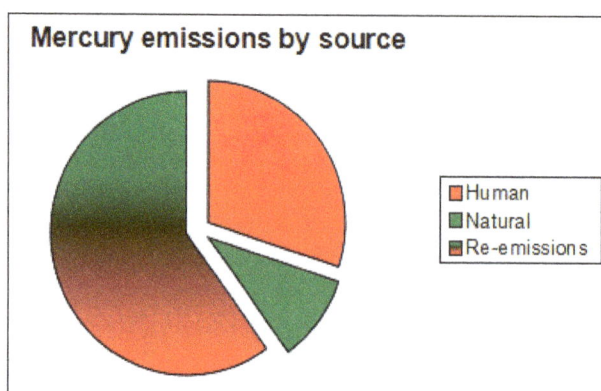

Emissions of mercury

Mercury emissions have decreased drastically in recent years, however small amounts of mercury can still be considered toxic in foods. There are limits placed on how much methylmercury can be contained within food before it is contaminated. For example, the limit for young children in Canada is 0.2 micrograms per kilogram of body weight per day. So if a child that weighs 20 kilograms ingests just 4 micrograms (an extremely small amount) in a day, they would be considered to have exceeded the maximum dose. An analogy: if a 1 gram paperclip were made entirely of methylmercury and you cut it into 250 000 pieces, only one of these pieces would be required for this toxic dose.

Composition

Mercury itself is an element, just like lead or arsenic, and is classified as a heavy metal. It has many interesting characteristics, some of which make it a very dangerous pollutant. In the atmosphere, elemental mercury can either converted into more toxic inorganic compounds in which oxidized mercury (Hg^{2+}) combines with other elements, or it can combine with carbon to form a worse contaminant known as methylmercury (CH_3Hg). These compounds may fall onto land or water through precipitation, or they may fall as dry particles and find their way into a lake or ocean.

Effects

Humans are exposed to mercury in 2 ways:

- Eating fish contaminated with organic methylmercury.

- Inhaling elemental mercury (Hg) or inorganic salts (Hg^{2+})

The first is the largest contributor to human effects. Methylmercury is toxic and can cause extremely adverse effects when consumed, which can happen when humans eat highly contaminated fish. Through the process of biomagnification the concentration of methylmercury within fishes increases as one goes higher in the food chain. This makes it dangerous to consume top and apex predators such as those listed in Figure, and especially dangerous for pregnant women and young children to do so.

Effects of methylmercury can result in complex neurological problems, especially in young children and babies, by affecting the brain and nervous system. Potential issues include: cerebral palsy, delayed onset of walking or talking, learning disabilities, tremors, irritability, impaired coordination and memory loss. Pregnant mothers especially should not eat larger fish as their baby is vulnerable to these chemicals that attack developing organs.

Mercury consumption can be lethal. The largest first case of this was seen in Japan in 1956 in which thousands of people were killed. Mercury dumped from a chemical plant bioaccumulated in the fish and thousands of people were in turn affected by eating them.

Ammonia

Ammonia (NH_3) is a highly reactive and soluble alkaline gas. It originates from both natural and anthropogenic sources, with the main source being agriculture, e.g. manures, slurries and fertiliser application.

Excess nitrogen can cause eutrophication and acidification effects on semi-natural ecosystems, which in turn can lead to species composition changes and other deleterious effects.

Ammonia comes from the breakdown and volatilisation of urea. Emissions and deposition vary spatially, with "emission hot-spots" associated with high-density intensive farming practices. Other agriculture-related emissions of ammonia include biomass burning or fertiliser manufacture. Ammonia is also emitted from a range of non-agricultural sources, such as catalytic converters in petrol cars, landfill sites, sewage works, composting of organic materials, combustion, industry and wild mammals and birds.

Emissions

At the turn of the 21st century, total ammonia emissions in the UK were estimated to be 283 kt N yr^{-1} with 228 kt coming from agricultural sources. In 2010 the agricultural sector was responsible for 89% of UK NH3 emissions. National NH_3 emissions in the UK are mapped at a 5 km grid resolution, using the AENEID model for agricultural sources, and at a 1 km or 5 km grid resolution for non-agricultural sources.

Emissions trends have mostly been downward since peak in late 1980s and early 1990s but have now flattened. As the climate warms, volatilisation of ammonia emissions will lead to a further rise in ammonia concentrations.

High emission areas with intensive dairy farming can be distinguished from low emission areas with extensive sheep and beef farming or "hot-spot" patterns associated with intensive pig and poultry farming. Emissions from agricultural sources vary temporally with agricultural practice. Seasonal variation is also associated with climate; volatilisation being highest when it is warmer. Some non-agricultural emission sources (e.g. seabird colonies) contribute only small amounts to the overall NH_3 emissions in the UK but, due to their location, are often the dominant emission source in remote and otherwise "clean" areas. Larger seabird colonies have been shown to emit similar amounts of NH_3 to large intensive poultry farms.

Atmospheric Interactions

Atmospheric ammonia has impacts on both local and international (transboundary) scales. In the atmosphere ammonia reacts with acid pollutants such as the products of SO_2 and NO_x emissions to produce fine ammonium (NH_4^+) containing aerosol. While the lifetime of NH_3 is relatively short (<10-100 km), NH_4^+ may be transferred much longer distances (100->1000 km). Hence NH_3 emissions contribute to international transboundary air pollutant issues addressed by the UNECE Convention on Long Range Transboundary Pollution.

In addition to the transboundary effects, NH_3 has substantial impacts at a local level: emissions occur at ground level in the rural environment and NH_3 is rapidly deposited. As a result some of the most acute problems of NH_3 deposition are for small relict nature reserves located in intensive agricultural landscapes.

Ammonia can be volatilised, emitted into the atmosphere when the surface concentration exceeds that of the surrounding air. Losses of NH_3 by volatilisation from the application of nitrogen (N) fertilisers range from negligible amounts to >50% of the applied fertiliser N, depending on fertiliser/manure type (e.g. urea higher volatilization rates than ammonium nitrate), application practice (e.g. injection, surface application) and environmental conditions. Solubility and dissolution processes primarily drive the magnitude of NH_3 emissions, higher in warm drying conditions and smaller in cool wet conditions.

Concentrations and Deposition

Ammonia concentrations are monitored across the UK (UK pollutant deposition), and show large spatial variability, reflecting a combination of the large number of ground level sources, primarily related to livestock farming, and the very reactive nature of gaseous NH_3. Concentrations of NH_3 range from 10 µg m^{-3} in areas of intensive livestock production, especially dairy and beef production, to 0.1 µg m^{-3} in the Scottish Highlands, especially in the north-west of Scotland and in the Hebrides.

These concentrations can be used to estimate deposition although deposition varies with ecosystem type and meteorology. Due to the varying affinity and compensation points of ammonia for different habitats, expressed in differences in mean deposition velocities, the rates of ammonia deposition vary greatly between habitat types.

Maps of concentrations and depositions across the UK are mapped using the FRAME model and calibrated using the measured NH_3 values at monitoring stations. This means that maps of NH_3 dry deposition need to be interpreted with care, noting whether they refer to inputs to specific habitat types (e.g. woodland, shrublands and croplands) or net dry deposition averaged over entire grid squares. For the purpose of assessing critical loads exceedance, deposition values for the relevant habitats need to be used, rather than grid averages.

Areas at risk from ammonia/nitrogen impacts include those close to point sources and areas within intensive agricultural regions which see elevated ammonia concentrations.

Effects

Effects of ammonia have been established from transect studies downwind of significant NH_3 sources and a field release. Ammonia can be taken up through the leaves via stomata, increasing the potential for nutrient N uptake. The consequences of foliar uptake and processing of an alkaline gas for cellular functions, appear to drive the deleterious effects of NH_3 on terrestrial plants. Alkalinity is also thought to be a key driver for NH_3 effects on epiphytic lichens. Atmospheric NH_3 also impacts as NH_4^+, when the NH_3 deposits to plant surfaces, dissolves and is washed into the soil where it can increase soil acidity and interfere with base cation uptake. Effects represent the combined effects of uptake through shoots as NH_3/NH_4^+ and roots as NH_4^+.

Negative effects on vegetation occur via direct toxicity, when uptake exceeds detoxification capacity and, via N accumulation, which increases the likelihood of detrimental interactions with other abiotic and biotic stressors. Ammonia can also enrich a system with nitrogen putting under-storey species at risk as they become shaded by the expansion of nitrophiles (N loving plants) that use the additional N to increase productivity and expand the over-storey. Nitrogen enrichment affects competition for resources, favouring fast growing, and tall species with rapid N assimilation rates. Mosses and lichens are most at risk, they have limited detoxification capacity relative to their uptake potential and a large surface area relative to mass.

Many lichen species are sensitive to even small increases in NH_3 concentrations above c. 1μg m^{-3}. Current evidence suggests that the absence of acidophytic lichens (lichens loving acid conditions) from twigs and trunks of acid-barked trees, growing in NH_3 rich environments, is due to NH_3 neutralizing the bark pH. Sheppard et al. found that monthly NH_3 concentrations > 20 μg m^{-3} decimated Cladonia portentosa populations in less than one year and that after three years the concentration had fallen to < 3 μg m^{-3}. Wet deposited NH_4^+ caused only restricted damage.

In mosses, NH_3 exposure can increase both the N and amino acid content of ectohydric pleurocarpous mosses. Elevations in N and amino acid content have been proposed as a well coupled indicator of NH_3-N deposition. Moss species differ with respect to their N uptake, and presumably their tolerance. Some Sphagnum (bog mosses) appear to be very sensitive, especially those that lack the red-orange pigments, carotenoids, that protect against oxidative stress. Overall dry deposited ammonia-N drives species composition change and reduces species cover and diversity, much faster than the same unit of N in wet deposition.

Attributing both specific effects in the field and indicators can be challenging because ammonia is a form of nitrogen which is an essential plant growth nutrient. In addition, some of the effects are difficult to separate from those caused by management, or lack of shading of the under-storey.

Effects on vegetation are:

- Eutrophication leading to changes in species assemblages; increase in N loving species (e.g. grasses) and species that can up regulate their carbon assimilation at the expense of species that are conservative in their N use.

- Shift in dominance from mosses, lichens and ericoids (heath species) towards grasses like Deschampsia flexuosa, Molinia caerulea and ruderal species, e.g. Chamerion angustifolium, Rumex acetosella, Rubus idaeus.

- Increased risk of frost damage in spring.

- Increased winter desiccation levels in Calluna and summer drought stress.

- Increase in N loving epiphytes, e.g. Xanthoria parietina, at the expense of epiphytes that prefer acid bark.

- Increased incidence of pest and pathogen attack, e.g. heather beetle outbreaks.

- Direct damage and death of sensitive species, e.g. lichens and mosses, Sphagnum, Pleurozium schreberi.

- Reduced root growth and mycorrhizal infection leading to reduced nutrient uptake, sensitivity to drought and nutrient imbalance with respect to N that is taken up via the foliage.

- Increase in soil pH follows acidification.

- Ammonia excess will lead to increases in nitrification and denitrification, contributing to greenhouse gas emissions.

Lead

Lead (Pb) is a naturally occurring heavy metal that is found in the Earth's crust. It is denser than most common materials, soft, malleable and has a low melting point. This capability to change easily the state from solid to liquid at atmospheric pressure is favouring its facility of being introduced as an air pollutant into the atmosphere.

How does Lead get into the Air?

Major sources of lead in the air are mining, metal manufacturing and piston-engine aircraft operating on leaded aviation fuel. Other sources of lead air pollution are industrial productions, waste incineration, recycling, mobilization of previously buried lead, utilities and lead-acid battery manufacturers.

Peak of lead emissions to the atmosphere had place during Industrial Revolution and with the usage of leaded gasoline during the last decades of the 20th century. Nowadays, high lead emissions still have place, especially in developing countries where industrial emissions arising from coal burning prevail.

What are the Environmental and Health Effects of Lead Pollution?

Lead has not as much influence on the environment as other pollutants, but it can have a noticeable impact on plants. Lead accumulate on soils for a long time (hundreds or even thousands of years) and also can combine with other metals to inhibit photosynthesis. At high lead pollution levels, plants growth and survival may suffer adverse effects and can cause neurological problems to vertebrates.

Lead air pollution health effects on humans usually are neurological effects in children and cardiovascular effects (high blood pressure and heart disease, for example) in adults.

Lead exposure on humans can be very harmful, involving almost every organ and system in the human body. Once it is inhaled, lead is placed on the bloodstream and bones driving to a possible case of lead poisoning.

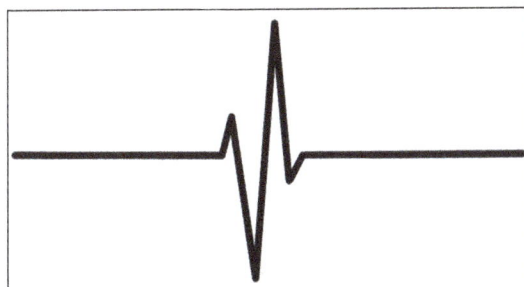

Lead Poisoning: Symptoms and Treatment

Apart from air pollution, lead poisoning come from different sources, such as water, dust, food or consumer products. The effects it may cause to each individual depends on the levels and the time exposure to the pollutant.

Some lead poisoning symptoms or signs that could indicate you are in danger are: headache, intermittent abdominal pain, loss of appetite, nausea, diarrhea, constipation, memory loss, kidney failure, male reproductive problems, depression, weakness, pain or tingling in the extremities and muscles.

In children, lead poisoning is prone to cause similar symptoms: loss of appetite, abdominal pain, vomiting, weight loss, constipation, anemia, kidney failure, irritability, lethargy, learning disabilities and behavioral problems. Others such as slow development of normal childhood behaviors (like the usage of words and talking) and permanent neurological problems (like learning deficits and lowered IQ) are also commonly diagnosticated to this segment. During the pregnancy, breathing lead polluted air may increase the risk of premature birth or low birth weight.

According to the level of poisoning and what part of your organism is affected, treatments may change. We will have a look to lead poisoning treatments due to air pollution.

When lead levels are high on blood, lead intoxication can be treated with chelation therapy or treatment of iron, calcium and zinc deficiencies, as these are treatments related to lead absorption.

How is Lead Air Pollution Controlled?

Over the years, developed countries have reached some agreements to control pollution. Each country or state has its own implementation plan. For example, the Clean Air Act made by the U.S. Environmental Protection Agency (EPA) establishes National Ambient Air Quality Standards (NAAQS) for those criteria pollutants considered most harmful for health and the environment. Another example is the European Union Air Quality Directive by the European Environmental Agency, which also establishes some standards and tips to reduce air pollution, both indoors and outdoors.

Lead is considered both a primary and a secondary criteria pollutant due to its effects either on public health and the environment. Maximum permitted or recommended levels of lead in the air are:

- NAAQS: cannot exceed 0,15 $\mu g/m^3$ for 3 months average.

- EU Air Quality Directive: cannot exceed 0,5 $\mu g/m^3$ over a year.

Chlorofluorocarbons

Chlorofluorocarbons (CFCs) are a group of compounds which contain the elements chlorine, fluorine and carbon. At room temperatures, they are usually colourless gases or liquids which evaporate easily. They are generally unreactive and stable, non-toxic and non-flammable. CFCs are also a part of the group of chemicals known as the volatile organic compounds (VOCs).

The properties of CFCs make them useful for a variety of commercial and industrial purposes: as a propellant in aerosol sprays (now banned in the US and Europe), in refrigeration and air conditioning systems, in foams, in cleaning solvents and in electrical components.

Most CFCs have been released to the atmosphere through the use of aerosols containing them and as leakages from refrigeration equipment. Other releases may occur from industry producing and using them and other products containing them. There are not thought to be any natural sources of CFCs to the environment.

CFCs are unlikely to have any direct impact on the environment in the immediate vicinity of their release. As VOCs, they may be slightly involved in reactions to produce ground level ozone, which can cause damage to plants and materials on a local scale. At a global level however, releases of CFCs have serious environmental consequences. Their long lifetimes in the atmosphere mean that some end up in the higher atmopshere (stratosphere) where they can destroy the ozone layer, thus reducing the protection it offers the earth from the sun's harmful UV rays. CFCs also contribute to Global Warming (through "the Greenhouse Effect"). Although the amounts emitted are relatively small, they have a powerful warming effect (a very high "Global Warming Potential").

How Might Exposure to it Affect Human Health?

Chlorofluorocarbons enter the body primarily by inhalation of air containing chlorofluorocabons, but can also enter by ingestion of contaminated water, or by dermal contact with chlorofluorocarbons. Inhalation of high levels of chlorofluorocarbons can affect the lungs, central nervous system, heart, liver and kidneys. Symptoms of exposure to chlorofluorocarbons can include drowsiness, slurred speech, disorientation, tingling sensations and weakness in the limbs. Exposure to extremely high levels of chlorofluorocarbons can result in death. Ingestion of chlorofluorocarbons can lead to nausea, irritation of the digestive tract and diarrhoea. Dermal contact with chlorofluorocarbons can cause skin irritation and dermatitis. Chlorofluorocarbons are involved in the destruction of the stratospheric ozone layer resulting in increased exposure to UV radiation which is known to cause skin cancer. The International Agency for Research on Cancer has not designated chlorofluorocarbons as a group in terms of their carcinogenicity. The International Agency for Research on Cancer has designated chlorofluoromethane and chlorodifluoromethane as being not classifiable as to their carcinogenicity to humans. However, exposure to chlorofluorocarbons at normal background levels is unlikely to have any adverse effect on human health.

Other Sources of Air Pollutants

Photochemical Oxidants

Photochemical oxidants are secondary pollutants that are formed during photochemical reactions in the atmosphere. These oxidants have short lifetimes but are continuously formed and destroyed through chemical reactions, leading to pseudo-steady-state concentrations that are important for chemical processing and can be inhaled. These oxidants include ozone, hydrogen peroxide, acids, peroxyacetyl nitrate, and reactive radicals. The reactive radicals, which include hydroxyl radical, oxygen radical, hydrogen radical, and several other radicals, have very short lifetimes and are not commonly measured. A large number of VOCs, SVOCs, and non-volatile organic compounds are also produced in photochemical smog, and some are oxidants. Ozone is often used as an indicator for these oxidant compounds.

Photochemical oxidants are formed in the presence of sunlight from the chemical reactions of VOCs and NO_x.

Given the nonlinear response of ozone production from the reaction of VOCs and NO_x, the relative source contributions to ozone cannot be directly scaled from the relative source contributions to VOCs and NO_x. Chemical transport models are needed to apportion the incremental ozone to sources.

Polycyclic Aromatic Hydrocarbons

Poor combustion conditions can lead to high emissions of PAHs and are often associated with liquid and solid fuel combustion. *Benzo[a]pyrene (B[a]P)* is a specific PAH formed mainly from the burning of organic material, such as wood, and from car exhaust fumes, especially from diesel vehicles. *B[a]P* pollution is predominantly a problem in countries where domestic coal and wood burning is common.

In 2007, it was estimated that the global total atmospheric emission of 16 PAHs came from residential/commercial biomass burning (60.5%), open-field biomass burning (agricultural waste burning, deforestation, and wildfire) (13.6%), and petroleum consumption by on-road motor vehicles (12.8%).

Other Organic Compounds

Thousands of organic compounds can be found in the atmosphere. They are components of fossil fuel, partially combusted components of fossil fuel, and pyrolysis products of fossil fuel; industrial chemical, food cooking, and biomass burning emissions; biogenic compounds emitted from plants; and organic compounds formed in the atmosphere. These compounds include VOCs, non-volatile organic compounds that are present in atmospheric PM, and SVOCs that are present in both the gas phase and the particle phase. Many known or suspected carcinogens come from combustion sources; they include benzene, 1,3-butadiene, formaldehyde, acetaldehyde, acrolein, and naphthalene. Industrial facilities and consumer products are also important sources of aromatic VOCs, oxygenated VOCs, and halogenated VOCs. These chemicals include benzene, toluene, xylenes, ethylbenzene, methyl ethyl ketone, acetophenone, and trichloroethylene. In addition, some VOCs of potential concern are also formed in the atmosphere from photochemical reactions; these include formaldehyde, acetaldehyde, and nitrobenzene. There is also a group of persistent organic pollutants (POPs), which include many SVOCs such as polychlorinated biphenyls, polybrominated biphenyls, furans, and dioxins, and several pesticides and insecticides that can be directly emitted from air pollution sources or re-emitted from previous contamination through volatilization or resuspension of soil material.

The three major sources of VOCs in Asia are stationary combustion, solvent and paint use, and transportation; the proportion of each of these sources varies between 25% and 50%, depending on the region. In Europe, solvent and product use was reported to contribute to about half of the total VOC emissions; the contributions of three other major sources of VOCs – commercial, institutional, and household energy use; road transportation; and energy production – were 10–20% each. In the USA, the relative source contribution reported in 2008 by the US EPA was 50% for transportation and 20% each for solvent use and industrial processes.

In recent years, significant progress in the development of emissions inventories has been made, including the current and future emissions of dioxins. To assess the sources of organic compounds that both are formed in the atmosphere and react in the atmosphere, such as formaldehyde, chemical transport models are needed. Several integrated assessments of emissions inventories of toxic organic compounds have been conducted and are used to provide an integrated risk from these sources by source and receptor.

Mineral Dust and Fibres

Resuspended dust from roadways, agricultural lands, industrial sources, construction sites, and deserts is a major source of PM in many regions of the world. Roadway dust also contains metals associated with motor vehicles. Agricultural soils often contain metals that accumulate from fertilizer and animal waste, and the content of dusts from industrial sources and construction sites will depend on the specific process activities occurring at those facilities.

Although fibres, such as asbestos, are not commonly measured in the outdoor atmosphere, they can be part of the atmospheric pollution mixture. The use of asbestos has been restricted or banned in many countries. However, outdoor air pollution with asbestos may still arise in some areas from releases from asbestos-containing building materials, asbestos brakes used on vehicles, and asbestos mining activity.

Bioaerosols

Bioaerosols are part of the atmospheric PM. The term "bioaerosol" refers to airborne biological particles, such as bacterial cells, fungal spores, viruses, and pollens, and their products, such as endotoxins. A wide range of these biological materials have been measured in the outdoor atmosphere, including moulds, spores, endotoxins, viruses, bacteria, proteins, and DNA. Knowledge about the dynamics and sources of airborne microbial populations is still scanty. Bioaerosols are believed to be ubiquitous, and studies demonstrate the long-range transport of microorganisms and biological particles in the atmosphere. Bioaerosols may derive from many sources, for example plants, suspension of soils containing biological materials, cooking, and burning of biological materials.

References

- Six-common-air-pollutants: ecolink.com, Retrieved 8 January, 2019

- Indoor-air-pollutants, indoor, healthy-air, our-initiatives: lung.org, Retrieved 13 May, 2019

- Toxic-air-pollutants, story, topics: 4cleanair.org, Retrieved 25 February, 2019

- Toxic-air-pollutants, air-pollution, outdoor, healthy-air, our-initiatives: lung.org, Retrieved 29 March, 2019

- Criteria-pollutants, air-pollutants, air-quality: idaho.gov, Retrieved 30 April, 2019

- Effects-car-pollutants-environment-23581: sciencing.com, Retrieved 29 June, 2019

- Particulate-matter-pm-basics, pm-pollution: epa.gov, Retrieved 16 January, 2019

- Overview-nox, pollutants, overview: apis.ac.uk, Retrieved 14 July, 2019

- Ground-level-ozone, common-contaminants, pollutants, air-pollution, services, environment-climate-change: canada.ca, Retrieved 11 January, 2019

- Volatile-organic-compounds-impact-indoor-air-quality, indoor-air-quality-iaq: epa.gov, Retrieved 18 April, 2019

- Lead, pollutants, air-pollution: airgo2.com, Retrieved 7 May, 2019

Chapter 4

Impacts of Air Pollution

Air pollution has a diverse range of ill effects which affect human health, domestic animals, environment and ozone layer. It can also lead to global warming, acid rain and wind erosion. The chapter closely examines these effects of air pollution to provide an extensive understanding of the subject.

Effects of Air Pollution on Human Health

The immediate effects of air pollution are hard to ignore. Watery eyes, coughing and difficulty breathing are acute and common reactions. An estimated 92 percent of the world's population live in areas with dangerous levels of air pollution and, even at seemingly imperceptible levels, air pollution can increase one's risk of cardiovascular and premature death.

Air pollution is almost as deadly as tobacco. In 2016, it was linked to the deaths of 6.1 million people. And it might harm you even before you take your first breath. Exposure to high levels of air pollution during pregnancy has been linked to miscarriages as well as premature birth, autism spectrum disorder and asthma in children.

Air pollution may damage children's brain development, and pneumonia, which kills almost 1 million children under the age of 5 every year, is associated with air pollution. Children who breathe in higher levels of pollutants also face a greater risk of short-term respiratory infections and lung damage.

Other conditions associated with high levels of air pollution include emphysema and chronic bronchitis, as well as lung cancer.

Pollutants can affect cardiovascular health by hardening the arteries and increase the risk of heart attack and strokes, and there is even emerging evidence that air pollution may be linked to mental health conditions and degenerative brain diseases such as Alzheimer's disease, Parkinson's disease and schizophrenia.

How Air Pollution Damages the Body

While air pollution's link to respiratory disease may seem obvious, its relationship to heart, brain and fetal health is less so. There are at least two possible mechanisms by which air pollution can harm parts of the body besides the nasal cavity and lungs.

The first has to do with inflammation, which is the body's way of repairing itself after an injury or illness. When the toxic soup of chemical particles and liquid droplets emitted by cars, power plants, fires and factories known as particulate matter is inhaled, the microscopic toxic dust can irritate nasal passages and result in an allergic-type response to the pollution, with symptoms like coughing and a runny nose.

Scientists believe that as the particles make their way deeper into the airways and into the lungs, the body may mistake it for an infection, triggering an inflammatory response. "When you have a bad head cold, you feel sick everywhere and your muscles might ache," Gerber said. "The same thing can happen when you breathe in pollution." Scientists also suspect that some toxic particles can escape the lungs and enter the bloodstream.

What Air Pollution does to the Body?

When air pollution enters the body, it can have both short-term and long-term effects on health. When particulate matter is inhaled, it can irritate tissue in the nasal cavity and cause coughing and a runny nose. Large particles can get stuck in the nose and sneezed out, but some scientists suspect that when finer particulate matter travels down the airways and into the lungs, it makes the body believe there is an infection. This could trigger inflammation, which scientists believe can cause shortness of breath and exacerbate pre-existing respiratory symptoms in which people, who have asthma or chronic obstructive pulmonary disease. Scientists hypothesize that this inflammation can spread from the lungs to other parts of the body, increasing risk of heart attack, stroke and other kinds of cardiovascular disease.

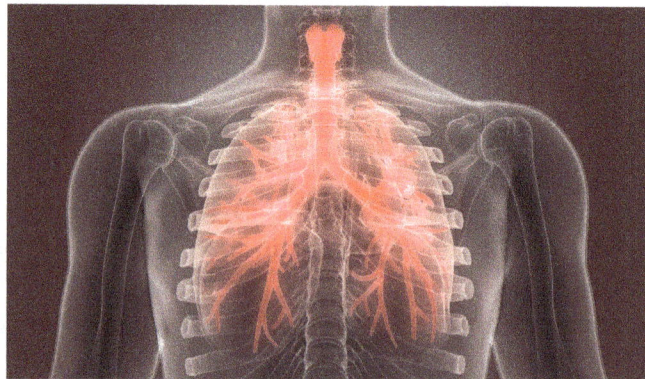

In pregnant ladies, air pollution may trigger inflammation throughout the body, including the uterus, which increases the risk of preterm birth.

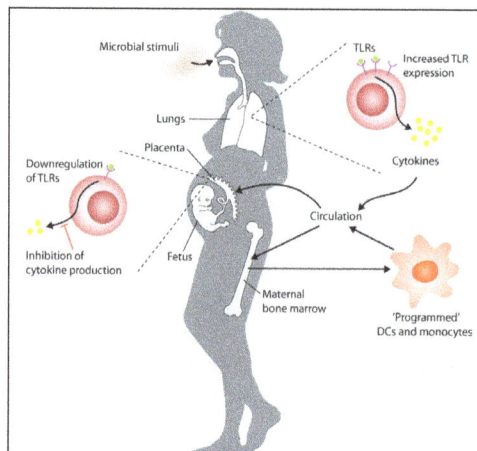

In children, air pollution has been linked to lung damage and can inhibit growth of lung function. It is unclear whether this permanently reduces lung function into adulthood.

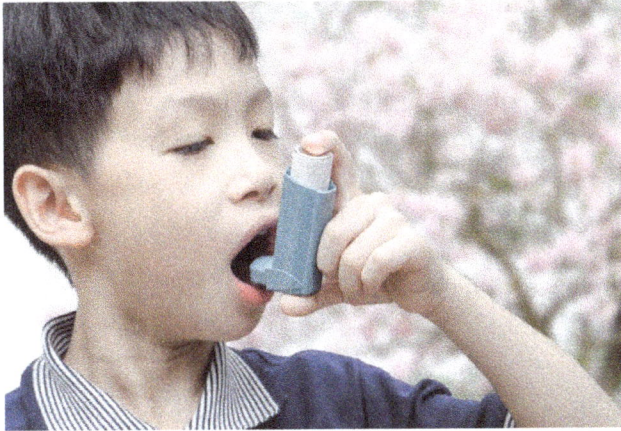

As particulate matter pollution goes down, life expectancy goes up. This may be because of particulate matter's effects on the heart and lungs. Globally three percent of heart and lung disease deaths, and five percent of lung cancer death, are linked to particulate matter.

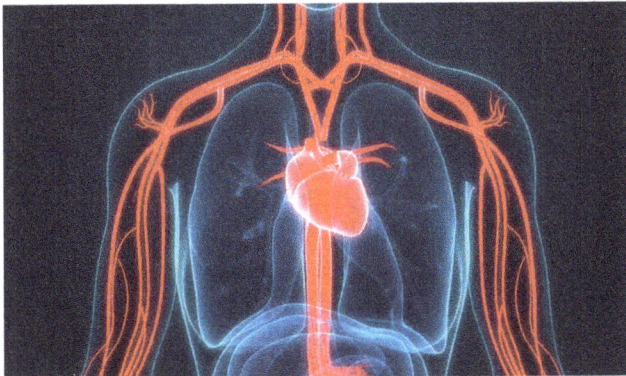

Effects on Domestic Animals

The effects of poor air quality on domestic animals principally can be divided in health damage caused by the in-door environment and by out-door air pollution. Pollutants may enter the system by inhalation or ingestion. In air pollution, mostly inhalation triggers the health problems, but occasionally deposition of particles from industrial exhaust on pasture land may affect health directly. Eventually, this may result in toxic residues in meat, milk or eggs without obvious clinical symptoms displayed by the animals producing these products. Problems with high dioxin levels in milk of dairy cows or zinc-induced arthritis in growing foals are examples of pasture grass contamination by deposits of smoke from nearby industrial activities.

The dog, the cat and the horse are exposed to the same health hazards as their masters regarding air pollution. Reineroa et al., reviewed the comparative aspects of feline asthma and brought evidence that important similarities between human and feline response to inhaled allergens exist.

The role environmental aeroallergens, however, was only shown in a few studies, but evidence suggests that some environmental allergens can cause disease in both cats and humans. the prevalence of asthma had increased over the last 20 years in cats in a large urban city. This seems to have happened in man as well.

Animals may be involuntarily acting as sentinels for detecting potential harmful effect on the organism of indoor air pollution. From the scope of comparative pathology, diseases of domestic animals associated with adverse environmental factors may give clues to the pathophysiology of the health disorders of man caused by air pollution.

Effects of Air Pollution on Animals

Production Animals

Pigs, poultry, cattle, goats and to a far lesser extend sheep are kept in indoor facilities for a variable part of their life, often for all of their life. For dairy cattle, goats and sheep these facilities are quite open and air quality is to a certain degree comparable with the outdoor air quality. The quality of this air is still much better than that of the closed facilities for swine and poultry. These buildings are rather closed and the natural or mechanically ventilation is via small air inlets and outlets. Indoor temperature is regulated to create optimal growing conditions, whereby heat loss via ventilation is kept to a level that is just on the boundary of what is still physiologically tolerable. The other reasons for closing these types of buildings as much as possible are the strict bio security procedures applied in order to avoid or reduce introduction of potential infectious material via air or fomites. The temperature in the facilities for optimal growth can be quite high. For instance, one day-old broiler chicks are kept at a room temperature of 34°C the first days of the raising period. Thereafter, ambient temperature will be lowered daily by 1 °C. The high temperatures facilitate growth of fungi and bacteria especially around the drinkers where water is spilled by the animals. The most common used litter for broilers is wood shavings. Sometimes alternatives such as shredded paper, chopped straw and pulverised bark or peat may be used. The bird's respiratory tracts are challenged by dust coming off the litter. Up to 40,000 broilers may be raised in a single house, on littered floors. A production cycle of broilers only takes 42 days on average. In this period the chicks will growth from about 60 grams to about 2000 grams. Thus, by the end of the raising period, the houses are well filled with animals and their activities increase dust levels in the air. In laying birds, although stocking density is lower, this beneficial effect on pollution, however, is offset by the longer housing period. The result is a larger accumulation of manure, usually in pits, which are only emptied infrequently. Hence, it is not surprising that especially in poultry houses high concentrations of ammonia, airborne dust, endotoxin and microorganisms can be measured.

Fattening pigs are kept in grid floored pens and thus are exposed to fumes of their own faeces and urine for their entire existence, which is of not more than 6-7 month. Also in many piggeries high levels ammonia, airborne dust, endotoxin and microorganisms can be found.

The indoor atmosphere in swine and poultry confinement buildings thus contains toxic gases, dusts and endotoxin in much higher concentrations than those in outdoor environments. Apart from minimal ventilation, poor stable design leading to poor homogeneity of ventilation causes locally stagnant air pockets., recommended maximal concentrations of gases or contaminants in piggeries are: 2.4 mg dust /m³; 7 ppm ammonia, 0.08 mg endotoxin/m³, 10^5 colony-forming units (cfu) of total microbes/m³; and 1,540 ppm, carbon dioxide. Concentrations of bacteria up to 1.1 x10⁶ cfu/m³, inhalable dust content of 0.26 mg/m³ and ammonia concentration of 27 ppm have been reported to occur in facilities during winter, while at summer lower concentrations were measured. Less difference between in- and outdoor temperature in summer allows better ventilation of the buildings.

A fraction of the smallest and most respirable particles are manure particles containing enteric bacteria and endotoxin. The concentration of these airborne bacteria and endotoxin, of course, is related to the level of pen cleanliness. Regarding generated toxic gasses, ammonia concentrations in the air are primarily affected by level of pen hygiene, but also by volume of the building, pig density and pig flow management. Furthermore, season plays a role as well as was shown by Scherer & Unshelm. Similar factors on ammonia levels are known to play a role in farrowing units and poultry houses. Ammonia is considered as one of the most important inhaled toxicant in agriculture. Dodd & Gross reported that 1000 ppm for less than 24 hour caused mucosal damage, impaired ciliary activity, and secondary infections in laboratory animals. Since this level is nearly never achieved, it is rather the long-term, low level exposure to ammonia that seems to be related to its ability to cause mucosal dysfunction with subsequent disrupting of innate immunity to inhaled pathogenic microorganisms. Generally, the toxic effects of chronic ammonia exposure do not extend into the lower respiratory tract.

In pigs this combined effects of ammonia and endotoxin predispose the animals to infections with viruses and bacteria, both primary pathogenic and opportunistic species. Although food producing animals appear to be capable of maintaining a high level of efficient growth in spite of marked degrees of respiratory disease, at a certain level of respiratory insufficiency rapid growth can no longer be attained. In that case the production results will be uneconomically. Ventilation is often at a just acceptable level. In their overview, Brockmeier et al.,summarized the facts on porcine respiratory diseases. They are the most important health problem for the industrial pork production today. Data collected from 1990 to 1994 revealed a 58% prevalence of pneumonia at slaughter in pigs kept in high-health herds. These animals originate from better farms and thus incidence of pneumonia in less well managed farms is higher. Respiratory disease in swine is mostly the result of a combination of primary and opportunistic infectious agents, whereby adverse environmental and management conditions are the triggers. Primary respiratory infectious agents can cause serious disease on their own, however, often uncomplicated infections are observed. More serious respiratory disease will occur if these primary infections become complicated with opportunistic bacteria. Common agents are porcine reproductive and respiratory syndrome virus (PRRSV), swine influenza virus (SIV), pseudorabies virus (PRV), possibly porcine respiratory coronavirus (PRCV) and porcine circovirus type 2 (PCV2) and Mycoplasma hyopneumoniae, Bordetella bronchiseptica, and Actinobacillus pleuropneumoniae. Pasteurella multocida, is the most common opportunistic bacteria, other common opportunists are Haemophilus parasuis, Streptococcus suis, Actinobacillus suis, and Arcanobacterium pyogenes.

Workers in pig or poultry facilities are exposed to the same increased levels of carbon monoxide, ammonia, hydrogen sulphide, or the dust particles from feed and manure as the animals. As a result, workers in swine production tend to have higher rates of asthma and respiratory symptoms than any other occupational group. Mc Donnell et al. studied Irish swine farm workers in concentrated animal feeding operations and measured their occupational exposure to various respiratory hazards. It appeared that swine workers were exposed to high concentrations of inhalable

(0.25–7.6 mg/m3) and respirable (0.01–3.4 mg/m3) swine dust and airborne endotoxin (166,660 EU/m3). Furthermore, the 8 hour time weighted average ammonia and peak carbon dioxide exposures ranged from 0.01–3 ppm and 430–4780 ppm, respectively.

Lesions caused by air pollution in production animals mainly include inflammatory processes. Neoplastic diseases are rather uncommon. This holds true for animals such as swine that are mainly kept indoors, as well as for cattle and sheep that are kept a variable part of their lives outdoors. This was shown in an abattoir survey some 5 decades ago performed in 100 abattoirs throughout Great Britain during one year. All tumours found in a total of 1.3 million cattle, 4.5 million sheep and 3.7 million pigs were recorded and histologically typed. Just 302 neoplasias were found in cattle, 107 in sheep and 133 in pigs. Lymphosarcoma was the commonest malignancy in all three species. Lymphosacoma was considered as entirely sporadic, since herds with multiple cases were not found in the UK. The other form, a lentivirus infection that causes outbreaks of enzootic bovine leukaemia was not present in the UK at those days. The 25 primary lung carcinomas in cattle were well-differentiated adenocarcinomas of acinar and papillary structure, squamous and oat-cell forms and several anaplastic carcinomas of polygonal-cell and pleomorphic types. They represented only 8.3 % of all neoplasms, occurring at a rate of 19 per million cattle slaughtered. No primary lung cancers were encountered in sheep or pigs.

Outdoor air pollution could affect farm animals kept at pastures in urban and peri-urban areas. In the past, a severe smog disaster in London was reported to have caused respiratory distress of prize cattle that were housed in the city for a cattle exhibition. It was likely the high level of sulphur dioxide that was responsible for acute bronchiolitis and the accompanying emphysema and right-sided heart failure. Since some of the city farms are located rather in the periphery of cities than in the centre, the inhaled concentrations of pollutants by production animals is likely less than the concentrations inhaled by pet animals living in the city centres or close to industrial estates.

Companion Animals

Bukowski & Wartenberg described clearly the importance of pathological findings in domestic animals with respect to analysis of the effects of indoor air pollution in a review. Radon and tobacco smoke are believed to be the most important respiratory indoor carcinogens. Already 42 years ago Ragland & Gorham reported that dogs in Philadelphia had an eight times higher risk developing tonsillar carcinoma than dogs from rural areas. Bladder cancer, mesothelioma, lung and nasal cancer in dogs are strongly associated with carcinogens released by human in-door activities. In cats, passive smoking increased the incidence of malignant lymphoma. By measuring urinary cotinine, passive smoking of the cats can be quantified. However, the late Catherine Vondráková (unpublished results) observed that there was no direct association with the amount of cigarettes that were smoked in a household and the level of cotinine in the urine of the family cat. Nevertheless, there was evidence that exposed cats showed reduced lung function. Measurement of lung function in small animals and in cats particularly, is difficult and usually only possible with whole body plethysmography. For this purpose the cat is placed in a Perspex plethysmography box. Whether this method has sufficient accuracy is still to be proven.

The effect of outdoor air pollution on companion animals, so far, has not been studied extensively. Catcott however described that in the smog incident of 1954 in Donora, Pennsylvania about 15%

of the cities dogs were reported to have experienced illness. A few died. Diseased dog were mostly less than 1 year old. Symptoms were mostly mild respiratory problems lasting for of 3-4 days. Also some cats had been reported ill. Further indirect evidence exists provided by observations made during the smog disaster of 1950 in Poza Rica Mexico. Many pets were reported ill or died. Especially canary birds appeared sensitive, since 100% of the population died. The cause of mortality in the dogs and cats, however, was not professionally established; the information was merely that what the owners had reported, when asked on the incident.

Dogs with chronic bronchitis and cats with airways inflammatory disease are at increased risk of exacerbating their conditions if exposed to prolonged urban air pollutants. In this respect they respond similar to man.

Horses

The reason for the domestication of the horse must be attributed to its athletic ability. The quieter donkey and the ox had been domesticated earlier as draft animals. The horse is one of the mammals with the highest relative oxygen uptake and therefore capable of covering long distances at high speed. The tidal volume of a 500 kg horse at rest is 6-7 L and at racing gallops 12-15 L. At rest a horse breathes 60-70 L of air per minute, which corresponds to about 100,000 L/day. During a race, the ventilation rate increases up to 1800 L /min. With this huge amount of air moving in and out the respiratory track, large quantities of dust particles are inhaled and may sediment in the airways. This on its term could have adverse consequences for lung function. Any decrease of lung function could affect the horse's performance over any distance that is longer than 400 meters. Respiratory problems have a direct impact on the racing career of racehorses, if not successfully treated. Horses that are submitted to less intensive exercise, however, can perform up to expectation for quite a long time, if they are only affected by a small decrease of lung function. This can easily be understood if one considers the huge capacity of the equine cardiopulmonary system.

Horses are not exposed to the negative effects of tobacco smoke or radiation, because stables and the living rooms of man mostly do not share common air spaces. Yet, this does not automatically imply that there is a healthy atmosphere in a horse stable. In those countries where horses are kept in stalls, sub acute and chronic respiratory diseases are serious and common problems. In countries like New Zealand, where horses live almost exclusively outdoors, these diseases are less well known.

Many equestrian enterprises are situated in the periphery of urbanized areas. Thus urban air pollution must be considered next to the health challenge by poor indoor air quality. In the suburban and urban enterprises mostly adult animals are employed. Riding schools, racehorse training yards and fiacre horse enterprises are examples of yards that may be located in or near city parks or urban green zones. Horses on these yards are either housed in barns or in individual open-fronted loose boxes. The latter have top doors that are mostly left open in order to optimize air circulation. Nevertheless, in many of these boxes due to their small doors, the minimal air change rate of 4/ hour is hardly attained.

The younger animals mainly are kept in rural areas, mostly at stud farms. Here they are kept outdoors partly or continuously. In winter and prior to horse auctions the youngsters will be stabled for longer periods, just to the moment that many of them will be shipped to suburban or urban enterprises. Other young animals will remain in the countryside. A special category of animals are the breeding animals. After having served in sporting events for short or longer periods in the (sub)urban environment, these animals return to the countryside. Mares are bred to stallions and are mostly kept at pasture for all day, or at least a part of the day. If housed, stables are not necessarily well designed and are as traditional as those of racehorses. Thus, exposure to poor air quality is not uncommon in broodmares. Breeding stallions, have only limited freedom, and yet remain large parts of the day in the barn. Stallion barns are mostly better designed than those for mares; often the more valuable stallions have open-front boxes.

Principally almost all horses will be exposed during a variable period of their life to air of poor quality. The sports and working horses stabled and exercised in (sub)urban regions are exposed to the air pollution caused by traffic and industrial activities too. Indoor and outdoor air pollution must have an impact on the lung health of our horses. Therefore it is not unexpected that respiratory disease is a major problem for horse industries worldwide.

Fiacre horses of Vienna waiting for tourist. Horses are daily exposed for at least 12 hours to air of the inner city. Most horses are also housed in the city.

Traditional stable design for horses is based on non-empirical recommendations extrapolated from studies of other agricultural species, ignoring fundamental differences in requirements of the equine athlete. Even now in 2010, only a fraction of the horses are housed in modern well designed stables. But even in the traditional stables, with a median floor space of about 12 m² stocking density is much less than with production animals. Moreover, many horses have their individual living area, but often still share a common airspace with poor air quality.

Organic dust in the common or individual air space, released by moving of bedding and hay is the main pollutant in horse stables Sometimes dust levels in stalls are less than 3 mg/m³, but during mucking out, the amount increased to 10-15 mg/m³, of which 20 - 60% is of respirable particles. Measured at the level of the breathing zone, during eating of hay, dust levels may be 20-fold higher than those measured in the stable corridor. Dust concentrations of 10 mg/m³ are known to be associated with a high prevalence of bronchitis in humans. Apart from hay and bedding, cereal food may contain considerable levels of dust. It has been shown that dry rolled grains may contain 30 – 60-fold more respirable dust than whole grains or grains mixed with molasses. Respirable dust is defined as particles smaller than 7 µm Respirable particles are capable of reaching the alveolar membrane and interact with alveolar cells and Clara cells. In this respect current findings by Snyder et al., in chemical and genetic mouse models of Clara cell and Clara cell secretory protein

(CCSP) deficiency coupled with Pseudomonas aeruginosa LPS elicited inflammation provide new understanding on the pathophysiology of chronic lung damage.

Kaup et al. mention that their ultrastructural study suggests that Clara cells are the main target for antigens and various mediators of inflammation during bronchial changes that occur in horses with recurrent airway obstruction (RAO).

The main constituents of stable dust are mould spores and it may contain at least 70 known species of fungi and Actinomycetes. Most of these micro organisms are not considered as primary pathogens. Occasionally infection of the guttural pouch with Aspergilles fumigatus may occur. The guttural pouch is a 300 mL diverticulum of the Eustachian tube.

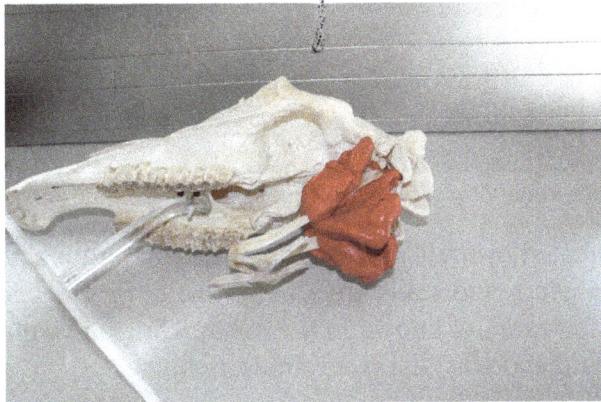

Horse skull with plastinated guttural pouches.

The walls of the guttural pouches are in contact with the base of the skull, some cranial nerves and the internal carotic artery. In case of a fungal infection of the air sac, the fungal plaque is commonly located at the dorsal roof, but may occupy the other walls as well. The fungus may invade and erode the wall of the adjacent artery. The resulting haemorrhage is not easily controlled and the horse may die due to blood loss.

A special infection associated with inhalation of bacteria present in the dust generated by dried faeces is the pneumonia caused by Rhodococcus equi of young foals. R.equi is a conditional pathogen causing disease in immunologically immature or immune-deficient horses. It can even cause disease in immuno compromised man. The key to the pathogenesis of R. equi pneumonia is the ability of the organism to survive and replicate within alveolar macrophages by inhibiting phagosome-lysosome fusion after phagocytosis. Only the virulent strains of R. equi having virulence-associated plasmid-encoded 15–17 kDa proteins (VapA) cause the disease in foals. This large plasmid is required for intracellular survival within macrophages. Next to VapA an antigenically related 20-kDa protein, VapB is known. These two proteins however are not expressed by the same R. equi isolate. Additional genes carring virulence plasmids e.g. VapC, -D and –E are known. These are co-ordinately regulated by temperature with VapA . Expression of the first occurs when R. equi is cultured at 37 °C, but not at 30° C. Thus it is plausible that the majority of cases of R. equi pneumonia are seen during the summer months. The prevalence of R. equi pneumonia is further associated with the airborne burden of virulent R. equi, but unexpectedly it seems not directly to be associated with the burden of virulent R. equi in the soil. Only under special conditions of the soil, the virulent organisms may be a thread to foals. Dry soil and little grass and holding pens and lanes which are sandy, dry, and lack sufficient grass cover are associated with elevated airborne concentrations of virulent R. equi. Hence, Muscatello et al. consider that environmental management strategies aiming to reduce the level of exposure of susceptible foals to airborne virulent R. equi likely will reduce the impact of R. equi pneumonia on endemically affected farms.

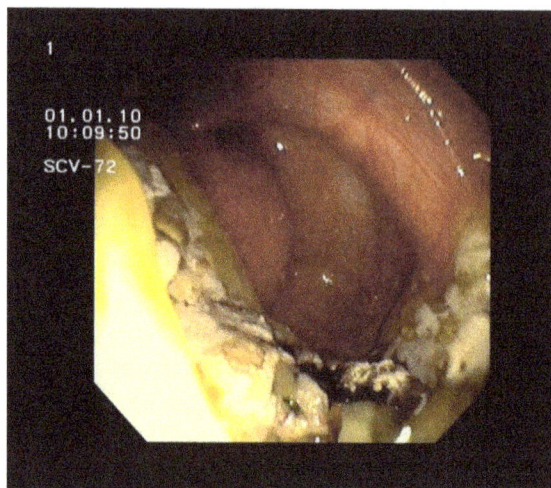

Endoscopic view of the guttural pouch with a mycotic plaque.

If contaminated dust is inhaled by foals of less than 5 month, pulmonary abscesses will develop. Faecal contamination of pasture and stalls are a prerequisite for the bacteria to establish. Other dust-born bacterial infections are not known in the horse. The non viable components of dust appear to play a major role in the airway diseases of mature horses.

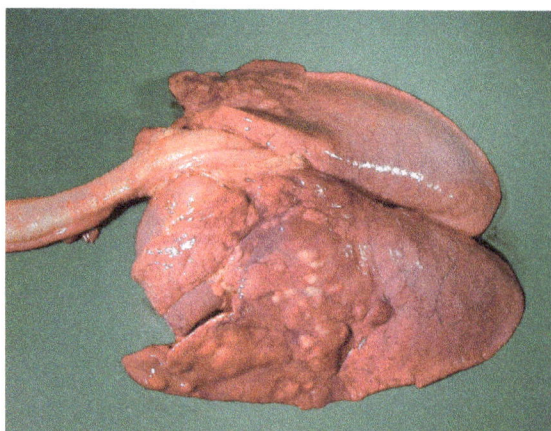

Pulmonary abscesses

Any threshold limiting value (TLV) for exposure to mould spores or dust is yet not known in horses. In man working for 40 h/week in a dusty environment, the TLV is 10 mg/m³. However, chronical exposure of 5 mg/m³ caused serious loss of pulmonary function in operators of grain elevators. Also Khan & Nachal, 2007 showed that long-term exposure to dust or endotoxin is important for the development of occupational pulmonary diseases in man. In this respect long periods of stabling causing a cumulative exposure effect of dust and endotoxins could result in the development of pulmonary disease in both horses that are susceptible to respiratory disorders and horses that are otherwise healthy.

Generally, horses that are exposed to excess organic dust will develop mild, often subclinical lower airway inflammation. This may contribute to poor performance. The symptoms initially seem to share common aspects with the organic dust toxic syndrome in man. Some horses could show severe hyperreactivity to organic dust and will display asthma-like attacks after exposure. Especially the feeding of mouldy hay is a well-known risk factor for this. Commonly incriminated allergens for such sensitive horses are the spores of Aspergillus fumigatus and endotoxins. The specific role of β-glucans is still in discussion.

The origin of the moulds may be found in the feedstuff offered to horses. Buckley et al. analysed Canadian and Irish forage, oats and commercially available equine concentrate feed and found pathogenic fungi and mycotoxins. The most notable fungal species were Aspergillus and Fusarium. Fifty per cent of Irish hay, 37% of haylage and 13% of Canadian hay contained pathogenic fungi. Apart from problems by inhalation, these fungi may produce mycotoxins that are rather ingested with the feed than inhaled. T2 and zearalenone appeared to be the most prominent. Twenty-one per cent of Irish hay and 16% of pelleted feed contained zearalenone, while 45% of oats and 54% of pelleted feed contained T2 toxins.

Next to fungal antigens, inhaled endotoxins induce a dose dependent airway inflammatory response in horses and even a systemic response on blood leucocytes can be observed. Inhaled endotoxins in horses suffering RAO are likely not the only determinants of disease severity, but do contribute to the induction of airway inflammation and dysfunction.

Whittaker et al. measured total dust and endotoxin concentrations in the breathing zone of horses in stables. Dust was collected for six hours with an IOM MultiDust Personal Sampler (SKC) positioned within the breathing zone of the horse and linked to a Sidekick sampling pump. The study confirmed earlier studies that forage has a greater effect on the total and respirable dust and endotoxin concentrations in the breathing zone of horses than the type of bedding.

Due to absence of slurry pits under their living area and the low stocking density, noxious gases generated indoors generally play a less important role in development of equine airway disease. Nevertheless, with poor stable hygiene, ammonia released from the urine by urease producing faecal bacteria may contribute to airway disease too.

The effect of air pollution on horses working in the open air has not been extensively studied, but the few studies performed on ozone showed that horses appear less susceptible to the acute effects of ozone compared to humans or laboratory. Marlin et al. 2001 found that the anti-oxidant activity of glutathione in the pulmonary lining fluid is likely a highly efficient protective mechanism in the horse. Although it is not likely that ozone is a significant risk factor for the development of respiratory disease in horses, the ability of ozone to act in an either additive or synergistic way with other agents or with already existing disease can not be neglected. Foster described that this occurs in humans. Diseases associated with poor air quality are follicular pharyngitis, equine inflammatory airway disease and recurrent airway obstruction.

In man exposed to air pollution in large cities, respirable particles and toxic gas levels appear to be associated with acute and subacute cardiopulmonary mortality. Such effects have not been noticed in horse exposed to urban air pollution.

Follicular Pharyngitis

Follicular pharyngitis in horses causes narrowing of the pharyngeal diameter and increased upper respiratory airway resistance with impairment of ventilation at high speeds. The symptoms are a snoring noise at in- and expiration during high-speed exercise. The disease is easily detected by endoscopy. The disease was previously attributed to a variety of viral infections, but according to Clarke et al. it must be considered as a multi factor disease. The disease is mostly self-limiting within a variable time interval.

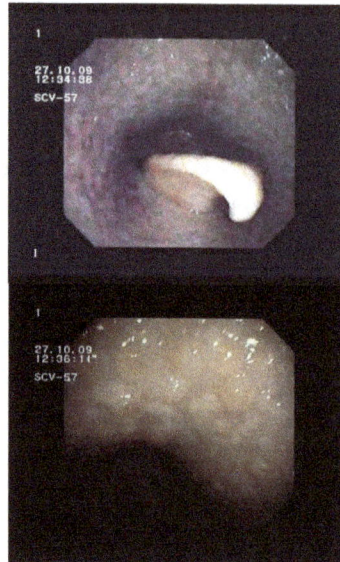

Endoscopic view of follicular pharyngitis - Upper panel: swollen pharynx and an oblique view on the side and apex of the epiglottis. Lower panel: a closer view of the large lymphoid follicle of the dorsal pharayngeal wall.

Chronic Bronchitis

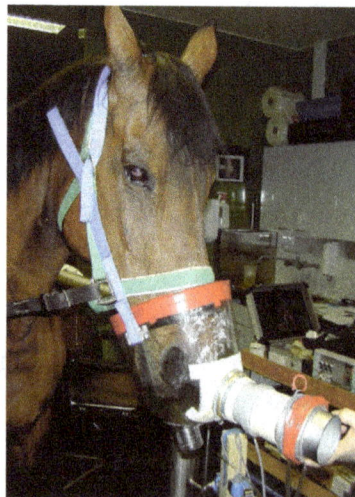

Intrapleural pressure and air flow measurement in a horse. Intrapleural pressure is measured via an oesophageal balloon connected to a plastic tube. Flow is measured with a Fleisch type pneumotachograph connected to an airtight facemask.

Cough and nasal discharge, caused by increased mucous production in the tracheo-bronchial tree, are common problems in equine medicine. It should be noticed that horses generally have a high threshold for coughing and thus cough is a strong indication for a respiratory disorder. In fact, coughing as clinical sign has 80% sensitivity for diagnosing tracheo-bronchial disorder. Today, endoscopy is the common technique to diagnose respiratory diseases. For this purpose, 3 meter long human colonoscopes are inserted via the nasal passages and the rima glottis into the trachea. The scope is further advanced into the larger bronchi. Via the endoscope samples can be taken. Commonly, a tracheo-bronchial aspirate or a broncho-alveolar lavage (BAL) is performed. Occasionally cytobrush samples or small biopsies are collected. The endoscopic image in relation to the cytological and bacteriological findings of the samples mostly leads to the diagnosis. The use of lung function tests in horses is only limited to those techniques that require little cooperation. Most commonly the intrapleural pressure in relation to airflow parameters is measured.

The two most important and frequent forms of bronchitis in the horse are Inflammatory Airway Disease (IAD) and Recurrent Airway Obstruction (RAO). In both conditions, a variable degree of airway hyperreactivity to inhaled dust particles plays a role. In the case of RAO, next to bronchiolar pathology, secondary changes in the larger airways and in the alveoli will develop.

Inflammatory Airway Disease (IAD)

IAD is a respiratory syndrome, commonly observed in young performance horses, but it is not exclusively a disease of the younger horse. Gerber et al. showed that many asymptomatic well-performing show-jumpers and dressage horse have signs of IAD. These horses are generally 7-14 years, which is older than the age of affected flat race horses that mostly is between 2 to 5 years.

Although a universally accepted definition of IAD does not exist, a working definition was proposed by the International Workshop on Equine Chronic Airway Disease. IAD is defined as a non-septic airway disease in younger, athletic horses that does not have a clearly defined aetiology. This approach was reconfirmed in the ACVIM Consensus Statement.

The incidence of IAD in thoroughbred and standard bred racehorses is estimated between 11.3 and 50%.

The clinical symptoms are often so subtle, that they may go unnoticed. In that case, disappointing racing performance may be the only indication for the presence of IAD. Endoscopic examination is the major help in diagnosing IAD. Mucous accumulation in the airways is commonly observed. The result of cytology of collected BAL fluid (BALF) samples is an important parameter for diagnosing the disease. Various inflammatory cells can be seen in cytospins of BALF samples. In contrast to RAO, slightly increased numbers of eosinophil granulocytes may be observed.

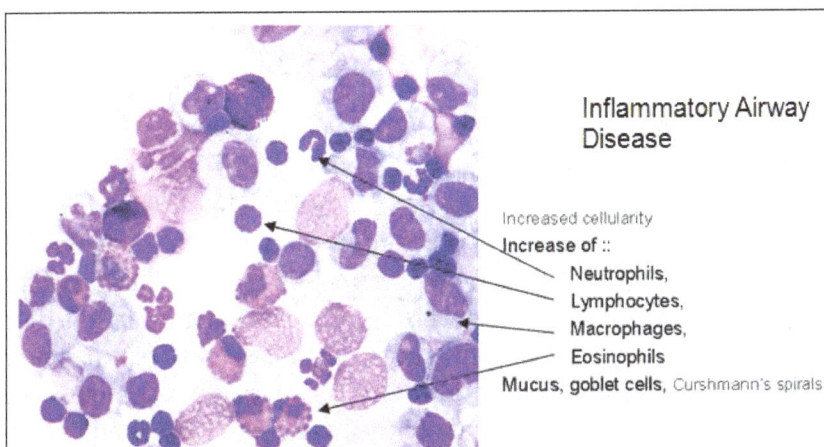

Cytospin of BALF of a horse with IAD. Romanowsky stain.

There is consensus that the clinical symptoms should include airway inflammation and lung dysfunction. However clinical signs are rather obscure and lung function test may only show very mild changes in respiratory resistance. At endoscopy the horses may have accumulated secretions in the trachea without necessarily displaying cough. Therefore, in contrast to other respiratory disorders, cough is an insensitive indicator of IAD in racehorses. IAD in racehorses seems to diminish with the time being in a training environment.

Respiratory virus infections do not appear to play a direct role in the syndrome, but there is still no consensus on their indirect role in the development of IAD. Bacterial colonisation of the respiratory mucosa is regularly detected. This could be associated with decreased mucociliary clearance. Poor mucosal clearance on its term could be the result of ciliar damage by dust or toxic gases such as ammonia. Common isolates include Streptococcus zooepidemicus, S. pneumoniae, members of the Pasteurellaceae (including Actinobacillus spp), and Bordatella bronchiseptica. Some studies have demonstrated a role for infections with Mycoplasma, particularly with M. felis and M. equirhinis.

It is estimated, however, that 35% to 58% of IAD cases are not caused by infections at all. Fine dust particles are assumed to be the trigger of these cases. Once IAD has established, long-term stay in conventional stables does not seem to worsen the IAD symptoms. Christley et al. reported that intense exercise, such as racing, may increase the risk of developing lower airway inflammation. Inhalation of dust particles from the track surface or of floating infectious agents may enter deep into the lower respiratory tract during hard exercise and cause impairment of pulmonary macrophage function together with altered peripheral lymphocyte function. In theory, intense exercise in cold weather may allow unconditioned air to gain access to the lower airways and cause airway damage, but studies in Scandinavia showed unequivocal results.

Many authors consider the barn or stable environment the important risk factor for development of respiratory disease in young horses. Interestingly, a study in Australia by Christley et al. reported that the risk of development of IAD decreased with the length of time horses were in training and thus stabled. An explanation for this finding is the development of tolerance to airborne irritants, a phenomenon that has been demonstrated in employees working in environments with high grain dust levels. IAD of the horse partly fit within the clinical picture of the human organic dust toxic syndrome (ODTS). Some evidence for this idea was presented by van den Hoven et al. et al., who could show inflammation of airways caused by nebulisation of Salmonella endotoxin.

Recurrent Airway Obstruction

Recurrent airway obstruction (RAO) is a common disease in horses. In the past, it used to be known as COPD, but as the pathophysiological mechanisms are more similar to human asthma than to human COPD, the disease is called RAO since 2001. The disease is not always clinically present, but after environmental challenge, horses show moderate to severe expiratory dyspnoea, next to nasal discharge and cough. Exacerbation of disease is caused by inhalation of environmental allergens, especially hay dust, that cause severe bronchospasm and in addition hypersecretion too. The mucosa becomes swollen while accumulated mucous secretions further contribute to airway narrowing. During remission, clinical symptoms may subside completely, but a residual inflammation of the airways and a hyperreactivity of the bronchi to nebulized histamine still remain present. A low degree of alveolar emphysema may develop as well, caused by frequent episodes of air trapping. In the past, severe end-stage emphysema was often diagnosed, but today this is rather uncommon and only sporadically occurs in old horses after many years of illness. The commonly accepted allergens that cause or provoke an exacerbation of RAO are especially spores of Aspergillus fumigatus and Fusarium spp.

Although the RAO share many similarities with human asthma, an accumulation of eosinophils in the BALF at exacerbation has never been reported. An asthma attack in humans is characterized by an early-phase response of bronchoconstriction, occurring within minutes of exposure to inhaled allergens. This phase is followed by a late asthmatic response with the continuation of airway obstruction and the development of airway inflammation. Mastcells play an important role in this early asthmatic response. The activation of mast cells after inhaling allergen results in the release of mastcell mediators, including histamine, tryptase, chymase, cysteinyl-leukotrines, and prostaglandin D2. These mediators induce airway smooth muscle contraction, clinically referred to as early-phase asthmatic response. Mastcells also release proinflammatory cytokines that, together with other mastcell mediators, have the potential to induce the influx of neutrophil and eosinophil granulocytes and the bronchoconstriction that are involved in the late-phase asthmatic response. Activation of other type of mastcell receptors can also induce mastcell degranulation or amplify the Fc-RI mediated mastcell activation.

In horses suffering RAO, such an early-phase response seems not to appear, whereas in healthy horses the early phase response does appear. This early-phase response may be a protective mechanism to decrease the dose of organic dust reaching the peripheral airways. Apparently in the horse with RAO, this protective mechanism has been lost and only the late-phase response will develop. The time of exposure to dust plays a determining role, as was shown by studies with exposure to hay and straw for 5 hours. This challenge caused an increase of histamine concentrations in BALF of RAO-affected horses, but not in control horses. In contrast, exposure of only 30 minutes to hay and straw did not result in a significant increase in BALF histamine concentration of RAO horses. A study of McPherson et al., 1979 showed that exposure to hay dust of at least 1 hour is needed to provoke signs. Also Giguère et al. and others provided evidence that the duration of exposure to organic dust must be longer than 1 hour. They are the opinion that the necessary exposure to provoke clinical signs of airway obstruction varies from hours to days in RAO affected horses.

The role of IgE-mediated events in RAO is still puzzling. Serum IgE levels against fungal spores in RAO horses were significantly higher than in healthy horses, but counts of IgE receptor-bearing cell in BALF were not significantly different between healthy and RAO affected horses. Lavoie et al. and Kim et al. held a T-helper cell response of type 2 responsible for the clinical signs, similar to human allergic asthma. However, their results are in contradiction with results of other research groups who could not find differences in lymphocyte cytokine expression patterns in cases with exacerbation of RAO compared to a control group.

The diagnosis of RAO is made if at least 2 of the following criteria are met: expiratory dyspnoea resulting in a maximal intra pleural pressure difference (ΔpPlmax) > 10 mm H_2O before provocation or > 15 mm H_2O after provocation with dust or by bad housing conditions. Any differential granulocyte count of > 10% in BALF is an indication for RAO. If symptoms can be ameliorated with bronchodilator treatment, the diagnosis is totally established. In some severe cases the arterial PaO_2 may be below 82 mmHg. After provocation with hay dust, RAO patients may reach equally low arterial oxygen levels too. Keeping the animals for 24 hours on pasture will quickly reduce clinical symptoms to a subclinical level.

The visible morphological changes are primarily located in the small airways and spread reactively to the alveoli and major air passages. Lesions may be focally, but functional changes may manifest themselves well throughout the bronchial tree. Bronchial lumina may contain a variable amount of exudate and may be plugged with debris. The epithelium is infiltrated with inflammatory cells, mainly neutrophil granulocytes. Furthermore, epithelial desquamation, necrosis, hyperplasia and non purulent peribronchial infiltrates may be seen. Fibrosing peribronchitis spreading in neighbouring alveolar septa was reported in severely diseased animals. The extent of these changes in the bronchioles is related to decrease of lung function, but changes may be distinctly focal in nature. Especially the function of Clara cells is important for the integrity of the bronchioles. Mildly diseased animals show loss of Clara cell granules next to goblet cell metaplasia even before inflammatory changes occur in the bronchioles. This together with the ultrastructural alterations found by Kaup et al. supports the idea of the damaging effects of dust and LPS. In severely affected horses Clara cells are replaced by highly vacuolated cells. Reactive lesions may be seen at the alveolar levels. These include necrosis of type I pneumocytes, alveolar fibrosis and variable degree of type II pneumocyte transformation. Furthermore, alveolar emphysema with an increase in Kohn's pores can be present. These structural changes may explain the loss of lung compliance in horses with severe RAO.

Whether there is any causal relation between RAO and IAD is not yet established. In both disorders, however, a poor climate in the stables plays a role. It could be theoretisized that IAD eventually may result in RAO, but Gerber et al. suggest there is no direct relation between IAD and RAO. In RAO the hyperreactivity induced by histamine nebulization or to air allergens is manifold more severe than in IAD, were only a mild bronchial hyperreactivity often can be shown.

Since long time, based on observations made on members of generations of horse families, it was believed that RAO has a hereditary component. Just recently Ramseyer et al. provided very strong evidence of an inherited predisposition to RAO on the basis of findings in two groups of horses. The same research group could demonstrate that mucin genes are likely to play a role too and that the IL4RA gene located on chromosome 13 is a candidate for RAO predisposition. The results gathered so far suggest that RAO seems to be a polygenic disease. Using segregation analysis for the hereditary aspects of the pulmonary health status for two stallion families, Gerber at al. showed that a major gene plays a role in RAO. The mode of inheritance in one family was autosomal dominant, whereas in the other horse family RAO seems to be inherited in an autosomal recessive mode.

Silicosis

Pulmonary silicosis results from inhalation of silicon dioxide (SiO_2) particulates. It is uncommon in horses; only in California a case series has been published. Affected horse showed chronic weight loss, exercise intolerance, and dyspnoea.

Effects of Air Pollution on the Environment

Human activities have increased the amount of pollutants introduced in the air that have direct and indirect effects in almost every ecosystem. Air pollution can lead to harmful consequences for living organisms through the inhalation of pollutants adverse weather conditions and others.

Effect on Plants and Vegetation

Air pollution has a lot of influence on vegetation by attacking its growth sources, such as airborne molecules, soil minerals or directly its organisms. Depending on the particular pollutant or environmental pollution conditions, main effects can be:

- Smog, as well as particulate matter high concentrations, reduces the amount of sun rays arriving to plants, denying or slowing plant growth. This kind of air pollution damages forests and crops, especially vegetables such as soybeans, wheat, tomatoes, peanuts and cotton.

- Ozone layer depletion increases the amount of UVB arriving to plants, and despite being prepared and adaptable to increasing levels of UVB, it can cause problems and modifications like form changes, nutrients distribution, developmental phases timing and secondary metabolism.

- Forest and plants can also be harmed by acid rain since it damages tree's leaves, robs the soil of essential nutrients and makes it hard for trees to take up water. All these issues imply growth and photosynthesis difficulties and more vulnerability to insects, diseases or bad weather. High concentrations of SO_x are also harmful for vegetation foliage and growth, and can contribute to formate acid rain. Ozone also produces similar symptoms, especially during the plants growing season.

- Lead can accumulate on soils for a long time (hundreds or even thousands of years) and by combination with other metals it can inhibit photosynthesis, what implies growth and survival issues for the surrounding vegetation.

- Nitrogen is essential for plants nutrition, but high levels of nitrogen dioxide or nitrogen monoxide pollution damage their lives.

Effects on Fauna and Flora

Air pollution may lead to harm an ecosystem as a whole and not only a particular organism from it. Some clear examples are:

- Marine ecosystems may experience high temperatures and exposure to UVB, reducing survival rate of phytoplankton and damaging early developmental stages of fish, shrimp, crab, amphibians and other marine animals. These effects are the result from the ozone layer depletion.

- Global warming is changing some ecosystems faster than the capability of animals and plants to adapt, leading to possible extinction of a huge amount of species. For example, ice sheets inhabited by polar bears are disappearing, warming oceans, more extreme weather conditions, etc.

- Due to the rising amount of carbon dioxide emissions and acid rain generation, the surface of oceans and water bodies has increased its acidity. This phenomenon is called ocean acidification and can lead to harmful consequences, such as depressing metabolic rates in jumbo squid, depressing the immune responses of blue mussels, and coral bleaching. Furthermore, ocean and lakes acidification makes water toxic to crayfish, clam, fish, and other aquatic animals. However, it can be good for some species as a trade-off, such as sea star, which increases its growth rate with highest water acid levels.

- Eutrophication, formed by phosphorus and nitrogen concentrations in water bodies, can even change the entire ecosystem from water to land (in extreme cases). Toxicity of the water, reduced amount of oxygen in deeper layers and difficult adaptability to the new substances may cause several damages into indigenous fauna and flora, leading to their reduction or even extinction.

Ozone Layer Depletion

Ozone depletion is the gradual thinning of Earth's ozone layer in the upper atmosphere caused by the release of chemical compounds containing gaseous chlorine or bromine from industry and other human activities. The thinning is most pronounced in the Polar Regions, especially over Antarctica. Ozone depletion is a major environmental problem because it increases the amount of ultraviolet (UV) radiation that reaches Earth's surface, which increases the rate of skin cancer, eye cataracts, and genetic and immune system damage. The Montreal Protocol, ratified in 1987, was the first of several comprehensive international agreements enacted to halt the production and use of ozone-depleting chemicals. As a result of continued international cooperation on this issue, the ozone layer is expected to recover over time.

Antarctic Ozone Hole

The most severe case of ozone depletion was first documented in 1985 in a paper by British Antarctic Survey (BAS) scientists Joseph C. Farman, Brian G. Gardiner, and Jonathan D. Shanklin. Beginning in the late 1970s, a large and rapid decrease in total ozone, often by more than 60 percent relative to the global average, has been observed in the springtime (September to November) over Antarctica. Farman and his colleagues first documented this phenomenon over their BAS station at Halley Bay, Antarctica. Their analyses attracted the attention of the scientific community, which found that these decreases in the total ozone column were greater than 50 percent compared with historical values observed by both ground-based and satellite techniques.

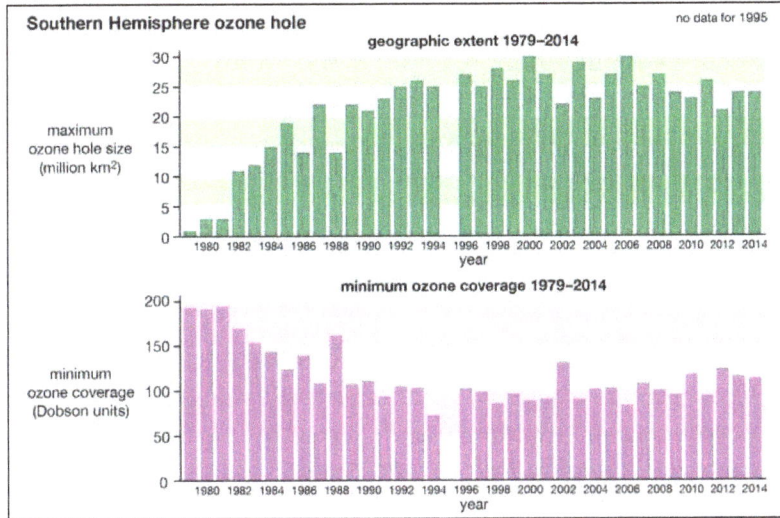

Southern Hemisphere ozone hole: Two bar graphs depicting the maximum ozone hole size and the minimum ozone coverage (in Dobson units) of the Southern Hemisphere ozone hole, 1979–2014.

As a result of the Farman paper, a number of hypotheses arose that attempted to explain the Antarctic "ozone hole." It was initially proposed that the ozone decrease might be explained by the chlorine catalytic cycle, in which single chlorine atoms and their compounds strip single oxygen atoms from ozone molecules. Since more ozone loss occurred than could be explained by the supply of reactive chlorine available in the Polar Regions by known processes at that time, other hypotheses arose. A special measurement campaign conducted by the National Aeronautics and Space Administration (NASA) and the National Oceanic and Atmospheric Administration (NOAA) in 1987, as well as later measurements, proved that chlorine and bromine chemistry were indeed responsible for the ozone hole, but for another reason: the hole appeared to be the product of chemical reactions occurring on particles that make up polar stratospheric clouds (PSCs) in the lower stratosphere.

During the winter the air over the Antarctic becomes extremely cold as a result of the lack of sunlight and a reduced mixing of lower stratospheric air over Antarctica with air outside the region. This reduced mixing is caused by the circumpolar vortex, also called the polar winter vortex. Bounded by a stratospheric jet of wind circulating between approximately 50° and 65 °S, the air over Antarctica and its adjacent seas is effectively isolated from air outside the region. The extremely cold temperatures inside the vortex lead to the formation of PSCs, which occur at altitudes of roughly 12 to 22 km (about 7 to 14 miles). Chemical reactions that take place on PSC particles convert less-reactive chlorine-containing molecules to more-reactive forms such as molecular chlorine (Cl_2) that accumulate during the polar night. (Bromine compounds and nitrogen oxides can also react with these cloud particles.) When day returns to Antarctica in the early spring, sunlight breaks the molecular chlorine into single chlorine atoms that can react with and destroy ozone. Ozone destruction continues until the breakup of the polar vortex, which usually takes place in November.

A polar winter vortex also forms in the Northern Hemisphere. However, in general, it is neither as strong nor as cold as the one that forms in the Antarctic. Although polar stratospheric clouds can form in the Arctic, they rarely last long enough for extensive decreases in ozone. Arctic ozone decreases of as much as 40 percent have been measured. This thinning typically occurs during years when lower-stratospheric temperatures in the Arctic vortex have been sufficiently low to lead to ozone-destruction processes similar to those found in the Antarctic ozone hole. As with Antarctica, large increases in concentrations in reactive chlorine have been measured in Arctic regions where high levels of ozone destruction occur.

Ozone Layer Recovery

The recognition of the dangers presented by chlorine and bromine to the ozone layer spawned an international effort to restrict the production and the use of CFCs and other halocarbons. The 1987 Montreal Protocol on Substances That Deplete the Ozone Layer began the phaseout of CFCs in 1993 and sought to achieve a 50 percent reduction in global consumption from 1986 levels by 1998. A series of amendments to the Montreal Protocol in the following years was designed to strengthen the controls on CFCs and other halocarbons. By 2005 the consumption of ozone-depleting chemicals controlled by the agreement had fallen by 90–95 percent in the countries that were parties to the protocol.

During the early 2000s, scientists expected that stratospheric ozone levels would continue to rise slowly over subsequent decades. Indeed, some scientists contended that, as levels of reactive chlorine and bromine declined in the stratosphere, the worst of ozone depletion would pass. Factoring in variations in air temperatures (which contribute to the size of ozone holes), scientists expected that continued reductions in chlorine loading would result in smaller ozone holes above Antarctica (which since 1992 have spanned more than 20.7 million square km [8 million square miles]) after 2040. The expected increases in ozone would be gradual primarily because of the long residence times of CFCs and other halocarbons in the atmosphere. Total ozone levels, as well as the distribution of ozone in the troposphere and stratosphere, would also depend on other changes in atmospheric composition—for example, changes in levels of carbon dioxide (which affects temperatures in both the troposphere and the stratosphere), methane (which affects the levels of reactive hydrogen oxides in the troposphere and stratosphere that can react with ozone), and nitrous oxide (which affects levels of nitrogen oxides in the stratosphere that can react with ozone).

Scientists in 2014 observed a small increase in stratospheric ozone—the first, they thought, in more than 20 years—which they attributed to worldwide compliance with international treaties regarding the phaseout of ozone-depleting chemicals and to upper stratospheric cooling because of increased carbon dioxide. Upon more thorough study, however, scientists in 2016 announced that stratospheric ozone concentrations had actually been increasing in the upper stratosphere since 2000 while the size of the Antarctic ozone hole had been decreasing. Overall zone concentrations away from the poles have continued to fall since 1998; however study showed that declines in the lower stratosphere outpaced gains in the upper stratosphere.

Since ozone is a greenhouse gas, the breakdown and anticipated recovery of the ozone layer affects Earth's climate. Scientific analyses show that the decrease in stratospheric ozone observed since the 1970s has produced a cooling effect—or, more accurately, that it has counteracted a small part of the warming that has resulted from rising concentrations of carbon dioxide and other greenhouse gases during this period. As the ozone layer slowly recovers in the coming decades, this cooling effect is expected to recede.

Visibility Effects of Air Pollution

Visibility is the ability to clearly see color and in distant views, can be impacted by air pollution. Many visitors come to parks to enjoy the spectacular vistas. Unfortunately, these vistas are sometimes obscured by haze, consisting of fine particles and gaseous air pollution in the atmosphere.

What Affects Visibility?

Air pollutants and wildfire smoke can reduce visibility at Denali National Park & Preserve, Alaska; clear to hazy views from left to right.

Fine particles and gaseous air pollution affect visibility to some degree in every national park. Air pollution can create a white or brown haze that affects how far we can see. Air pollution also affects how well we are able to see the colors, forms, and textures of a scenic vista. Haze results from air pollutants, such as fine particles that absorb and scatter sunlight. Haze is mostly caused by air pollution from industry and motor vehicles. Some haze can also occur naturally due to dust, fog, and wildfire smoke.

Processes

How is Visibility Impacted by Pollution?

Visibility is affected by the physical interaction of light with particles and gases in the atmosphere. However, visibility involves more than how light is absorbed and scattered by the atmosphere. Visibility is the process of perceiving the environment through the use of the eye-brain system.

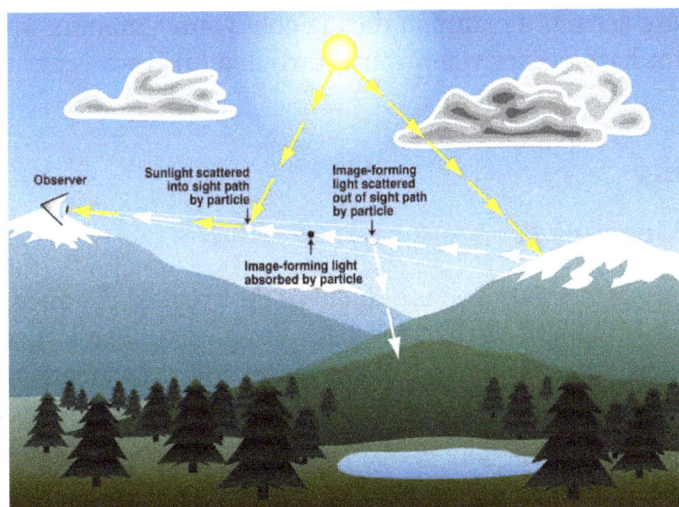

Particles in the air can impact the ability to see scenic vistas by scattering and/or absorbing image-forming light.

Important factors involved in seeing a scenic vista are shown in the following figure. Image-forming information from an object is reduced (scattered and absorbed) as it passes through the atmosphere to the eye. Sunlight is also added to the sight path by scattering processes. Sunlight and light reflected from the ground are absorbed and scattered by particles located in the sight path. Some of this scattered light remains in the sight path, and at times can become so bright that the image essentially disappears.

Types of Haze

What does Haze Look Like?

Types of haze include plumes, layered haze, and uniform haze.

Air pollution does not impact views on clear days but can be seen as a plume, layered haze, or uniform haze when air pollution is present. A plume of air pollution is a tight, vertically constrained layer of air pollution coming from a point source (such as a smoke stack). Layered haze is any confined layer of pollutants that creates as a visible contrast between that layer and the sky or landscape behind it. In an unstable atmosphere, plumes and layers mix with the surrounding atmosphere creating a uniform haze or overall reduction in air clarity.

Plumes and layered haze are more common during cold winter months when the atmosphere is more stagnant. Uniform haze occurs when warm turbulent air causes atmospheric pollutants to become well mixed.

Measuring Visibility

How is Visibility Measured?

Visual range is a measure of visibility and is defined as the greatest distance at which a large black object can be seen and recognized against the background sky. The larger the visual range the better the visibility. It is not directly measured but rather calculated from a measurement of light extinction which includes the scattering and absorption of light by particles and gases. Scattering is measured with nephelometers. Extinction depends on the mass and chemical composition of the particles and gases and is a quantitative measure of how the passage of light from a scenic feature to an observer is affected by air pollutants. Light extinction is reconstructed from measurements of particle mass and chemical composition.

Air quality monitoring station at Great Smoky Mountains NP in North Carolina and Tennessee.

Global Warming

Global warming is the phenomenon of increasing average air temperatures near the surface of Earth over the past one to two centuries. Climate scientists have since the mid-20th century gathered detailed observations of various weather phenomena (such as temperatures, precipitation, and storms) and of related influences on climate (such as ocean currents and the atmosphere's chemical composition). These data indicate that Earth's climate has changed over almost every conceivable timescale since the beginning of geologic time and that the influence of human activities since at least the beginning of the Industrial Revolution has been deeply woven into the very fabric of climate change.

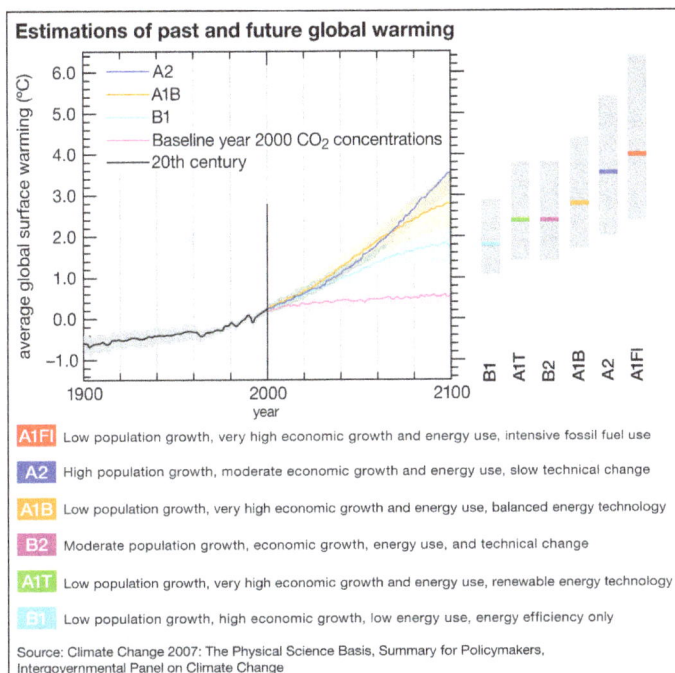

Global warming scenarios.

Giving voice to a growing conviction of most of the scientific community, the Intergovernmental Panel on Climate Change (IPCC) was formed in 1988 by the World Meteorological Organization (WMO) and the United Nations Environment Program (UNEP). In 2013 the IPCC reported that the interval between 1880 and 2012 saw an increase in global average surface temperature of approximately 0.9 °C (1.5 °F). The increase is closer to 1.1 °C (2.0 °F) when measured relative to the preindustrial (i.e., 1750–1800) mean temperature.

Graph of the predicted increase in Earth's average surface temperature according to a series of climate change scenarios that assume different levels of economic development, population growth, and fossil fuel use. The assumptions made by each scenario are given at the bottom of the graph.

- A1FI-low population growth, very high economic growth and energy use, intensive fossil fuel use.

- A2-high population growth, moderate economic growth and energy use, slow technical change.

- A1B-low population growth, very high economic growth and energy use, balanced energy technology.

- B2-moderate population growth, economic growth, energy use, and technical change.

- A1T-low population growth, very high economic growth, and energy use, renewable energy technology.

- B1-Low population growth, high economic growth, low energy use, energy efficiency only.

A special report produced by the IPCC in 2018 honed this estimate further, noting that human beings and human activities have been responsible for a worldwide average temperature increase of between 0.8 and 1.2 °C (1.4 and 2.2 °F) of global warming since preindustrial times, and most of the warming observed over the second half of the 20th century could be attributed to human activities. It predicted that the global mean surface temperature would increase between 3 and 4 °C (5.4 and 7.2 °F) by 2100 relative to the 1986–2005 average should carbon emissions continue at their current rate. The predicted rise in temperature was based on a range of possible scenarios that accounted for future greenhouse gas emissions and mitigation (severity reduction) measures and on uncertainties in the model projections. Some of the main uncertainties include the precise role of feedback processes and the impacts of industrial pollutants known as aerosols, which may offset some warming.

Many climate scientists agree that significant societal, economic, and ecological damage would result if global average temperatures rose by more than 2 °C (3.6 °F) in such a short time. Such damage would include increased extinction of many plant and animal species, shifts in patterns of agriculture, and rising sea levels. By 2015 all but a few national governments had begun the process of instituting carbon reduction plans as part of the Paris Agreement, a treaty designed to help countries keep global warming to 1.5 °C (2.7 °F) above preindustrial levels in order to avoid the worst of the predicted effects. Authors of a special report published by the IPCC in 2018 noted that should carbon emissions continue at their present rate, the increase in average near-surface air temperatures would reach 1.5 °C sometime between 2030 and 2052. Past IPCC assessments reported that the global average sea level rose by some 19–21 cm (7.5–8.3 inches) between 1901 and 2010 and that sea levels rose faster in the second half of the 20th century than in the first half. It also predicted, again depending on a wide range of scenarios, that the global average sea level would rise 26–77 cm (10.2–30.3 inches) relative to the 1986–2005 average by 2100 for global warming of 1.5 °C, an average of 10 cm (3.9 inches) less than what would be expected if warming rose to 2 °C (3.6 °F) above preindustrial levels.

The scenarios referred to above depend mainly on future concentrations of certain trace gases, called greenhouse gases, that have been injected into the lower atmosphere in increasing amounts through the burning of fossil fuels for industry, transportation, and residential uses. Modern global warming is the result of an increase in magnitude of the so-called greenhouse effect, a warming of Earth's surface and lower atmosphere caused by the presence of water vapour, carbon dioxide, methane, nitrous oxides, and other greenhouse gases. In 2014 the IPCC reported that concentrations of carbon dioxide, methane, and nitrous oxides in the atmosphere surpassed those found in ice cores dating back 800,000 years.

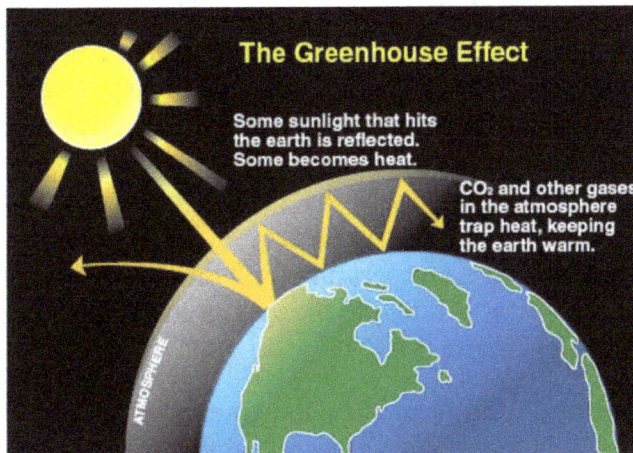

Greenhouse effect on Earth.

The greenhouse effect on Earth. Some incoming sunlight is reflected by Earth's atmosphere and surface, but most is absorbed by the surface, which is warmed. Infrared (IR) radiation is then emitted from the surface. Some IR radiation escapes to space, but some is absorbed by the atmosphere's greenhouse gases (especially water vapour, carbon dioxide, and methane) and reradiated in all directions, some to space and some back toward the surface, where it further warms the surface and the lower atmosphere.

Of all these gases, carbon dioxide is the most important, both for its role in the greenhouse effect and for its role in the human economy. It has been estimated that, at the beginning of the industrial age in the mid-18th century, carbon dioxide concentrations in the atmosphere were roughly 280 parts per million (ppm). By the middle of 2018 they had risen to 406 ppm, and, if fossil fuels continue to be burned at current rates, they are projected to reach 550 ppm by the mid-21st century—essentially, a doubling of carbon dioxide concentrations in 300 years.

A vigorous debate is in progress over the extent and seriousness of rising surface temperatures, the effects of past and future warming on human life, and the need for action to reduce future warming and deal with its consequences.

Causes of Global Warming

The Greenhouse Effect

The average surface temperature of Earth is maintained by a balance of various forms of solar and terrestrial radiation. Solar radiation is often called "shortwave" radiation because the frequencies of the radiation are relatively high and the wavelengths relatively short—close to the visible portion of the electromagnetic spectrum. Terrestrial radiation, on the other hand, is often called "longwave" radiation because the frequencies are relatively low and the wavelengths relatively long—somewhere in the infrared part of the spectrum. Downward-moving solar energy is typically measured in watts per square metre. The energy of the total incoming solar radiation at the top of Earth's atmosphere (the so-called "solar constant") amounts roughly to 1,366 watts per square metre annually. Adjusting for the fact that only one-half of the planet's surface receives solar radiation at any given time, the average surface insolation is 342 watts per square metre annually.

The amount of solar radiation absorbed by Earth's surface is only a small fraction of the total solar radiation entering the atmosphere. For every 100 units of incoming solar radiation, roughly 30 units are reflected back to space by either clouds, the atmosphere, or reflective regions of Earth's surface. This reflective capacity is referred to as Earth's planetary albedo, and it need not remain fixed over time, since the spatial extent and distribution of reflective formations, such as clouds and ice cover, can change. The 70 units of solar radiation that are not reflected may be absorbed by the atmosphere, clouds, or the surface. In the absence of further complications, in order to maintain thermodynamic equilibrium, Earth's surface and atmosphere must radiate these same 70 units back to space. Earth's surface temperature (and that of the lower layer of the atmosphere essentially in contact with the surface) is tied to the magnitude of this emission of outgoing radiation according to the Stefan-Boltzmann law.

Earth's energy budget is further complicated by the greenhouse effect. Trace gases with certain chemical properties—the so-called greenhouse gases, mainly carbon dioxide (CO_2), methane (CH_4), and nitrous oxide (N_2O)—absorb some of the infrared radiation produced by Earth's surface. Because of this absorption, some fraction of the original 70 units does not directly escape to space. Because greenhouse gases emit the same amount of radiation they absorb and because this radiation is emitted equally in all directions (that is, as much downward as upward), the net effect of absorption by greenhouse gases is to increase the total amount of radiation emitted downward toward Earth's surface and lower atmosphere. To maintain equilibrium, Earth's surface and lower atmosphere must emit more radiation than the original 70 units. Consequently, the surface temperature must be higher. This process is not quite the same as that which governs a true greenhouse, but the end effect is similar. The presence of greenhouse gases in the atmosphere leads to a warming of the surface and lower part of the atmosphere (and a cooling higher up in the atmosphere) relative to what would be expected in the absence of greenhouse gases.

It is essential to distinguish the "natural," or background, greenhouse effect from the "enhanced" greenhouse effect associated with human activity. The natural greenhouse effect is associated with surface warming properties of natural constituents of Earth's atmosphere, especially water vapour, carbon dioxide, and methane. The existence of this effect is accepted by all scientists. Indeed, in its absence, Earth's average temperature would be approximately 33 °C (59 °F) colder than today, and Earth would be a frozen and likely uninhabitable planet. What has been subject to controversy is the so-called enhanced greenhouse effect, which is associated with increased concentrations of greenhouse gases caused by human activity. In particular, the burning of fossil fuels raises the concentrations of the major greenhouse gases in the atmosphere, and these higher concentrations have the potential to warm the atmosphere by several degrees.

Radiative Forcing

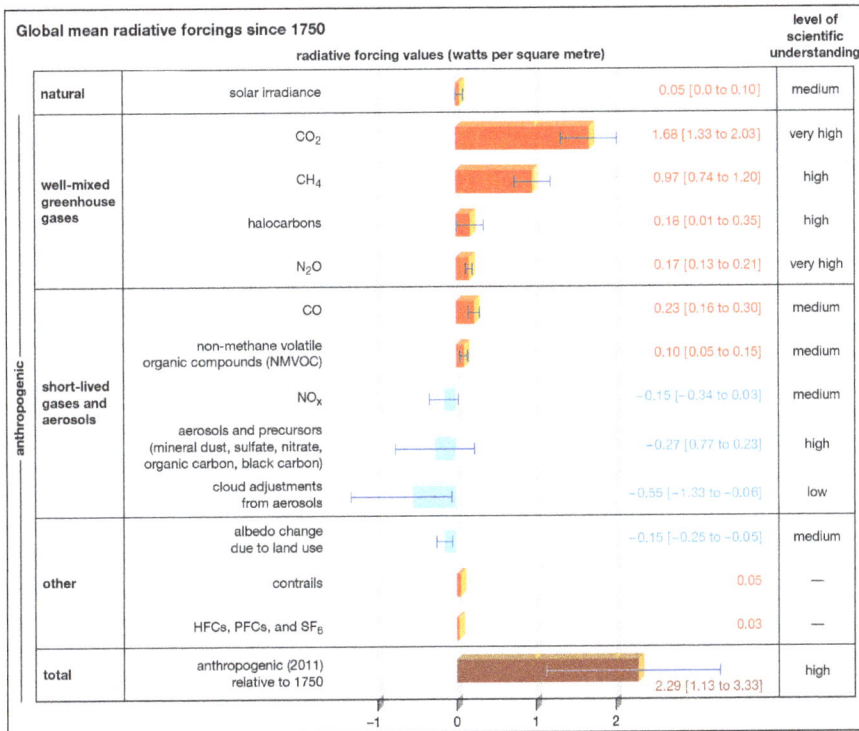

Global mean radiative forcings since 1750		radiative forcing values (watts per square metre)		level of scientific understanding
natural	solar irradiance		0.05 [0.0 to 0.10]	medium
well-mixed greenhouse gases	CO$_2$		1.68 [1.33 to 2.03]	very high
	CH$_4$		0.97 [0.74 to 1.20]	high
	halocarbons		0.18 [0.01 to 0.35]	high
	N$_2$O		0.17 [0.13 to 0.21]	very high
short-lived gases and aerosols	CO		0.23 [0.16 to 0.30]	medium
	non-methane volatile organic compounds (NMVOC)		0.10 [0.05 to 0.15]	medium
	NO$_x$		−0.15 [−0.34 to 0.03]	medium
	aerosols and precursors (mineral dust, sulfate, nitrate, organic carbon, black carbon)		−0.27 [0.77 to 0.23]	high
	cloud adjustments from aerosols		−0.55 [−1.33 to −0.06]	low
other	albedo change due to land use		−0.15 [−0.25 to −0.05]	medium
	contrails		0.05	—
	HFCs, PFCs, and SF$_6$		0.03	—
total	anthropogenic (2011) relative to 1750		2.29 [1.13 to 3.33]	high

Since 1750 the concentration of carbon dioxide and other greenhouse gases has increased in Earth's atmosphere. As a result of these and other factors, Earth's atmosphere retains more heat than in the past.

It is apparent that the temperature of Earth's surface and lower atmosphere may be modified in three ways: (1) through a net increase in the solar radiation entering at the top of Earth's atmosphere, (2) through a change in the fraction of the radiation reaching the surface, and (3) through a change in the concentration of greenhouse gases in the atmosphere. In each case the changes can be thought of in terms of "radiative forcing." As defined by the IPCC, radiative forcing is a measure of the influence a given climatic factor has on the amount of downward-directed radiant energy impinging upon Earth's surface. Climatic factors are divided between those caused primarily by human activity (such as greenhouse gas emissions and aerosol emissions) and those caused by natural forces (such as solar irradiance); then, for each factor, so-called forcing values are calculated for the time period between 1750 and the present day. "Positive forcing" is exerted by climatic factors that contribute to the warming of Earth's surface, whereas "negative forcing" is exerted by factors that cool Earth's surface.

On average, about 342 watts of solar radiation strike each square metre of Earth's surface per year, and this quantity can in turn be related to a rise or fall in Earth's surface temperature. Temperatures at the surface may also rise or fall through a change in the distribution of terrestrial radiation (that is, radiation emitted by Earth) within the atmosphere. In some cases, radiative forcing has a natural origin, such as during explosive eruptions from volcanoes where vented gases and ash block some portion of solar radiation from the surface. In other cases, radiative forcing has an anthropogenic, or exclusively human, origin. For example, anthropogenic increases in carbon dioxide, methane, and nitrous oxide are estimated to account for 2.3 watts per square metre of positive radiative forcing. When all values of positive and negative radiative forcing are taken together and all interactions between climatic factors are accounted for, the total net increase in surface radiation due to human activities since the beginning of the Industrial Revolution is 1.6 watts per square metre.

The Influences of Human Activity on Climate

Human activity has influenced global surface temperatures by changing the radiative balance governing the Earth on various timescales and at varying spatial scales. The most profound and well-known anthropogenic influence is the elevation of concentrations of greenhouse gases in the atmosphere. Humans also influence climate by changing the concentrations of aerosols and ozone and by modifying the land cover of Earth's surface.

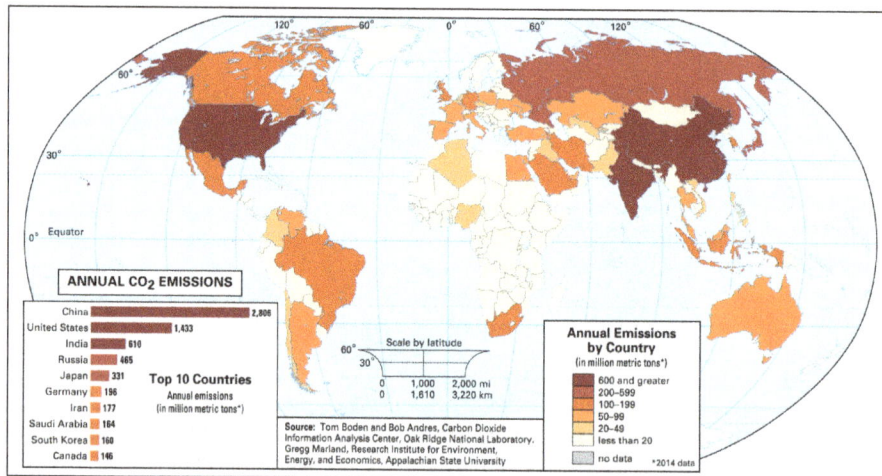

Carbon dioxide Emmisions.

Greenhouse Gases

Factories that burn fossil fuels help to cause global warming.

As discussed above, greenhouse gases warm Earth's surface by increasing the net downward longwave radiation reaching the surface. The relationship between atmospheric concentration of greenhouse gases and the associated positive radiative forcing of the surface is different for each gas. A complicated relationship exists between the chemical properties of each greenhouse gas and the relative amount of longwave radiation that each can absorb. What follows is a discussion of the radiative behaviour of each major greenhouse gas.

Water Vapour

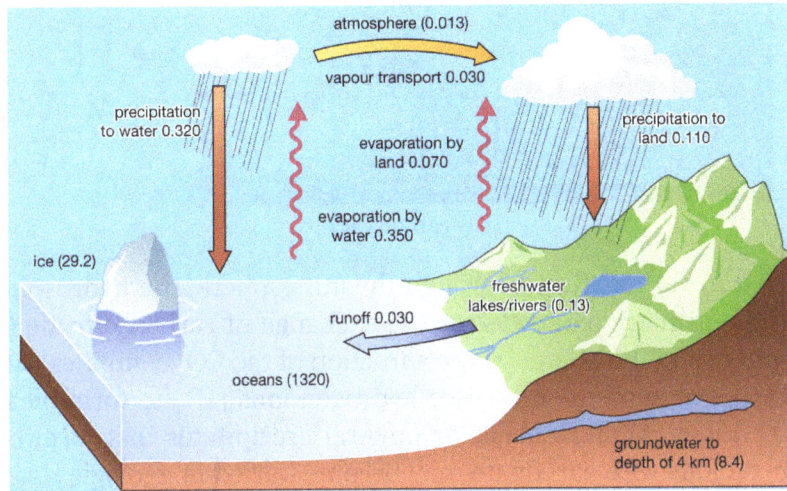

The present-day surface hydrologic cycle, in which water is transferred from the oceans through the atmosphere to the continents and back to the oceans over and beneath the land surface. The values in parentheses following the various forms of water (e.g., ice) refer to volumes in millions of cubic kilometres; those following the processes (e.g., precipitation) refer to their fluxes in millions of cubic kilometres of water per year.

Water vapour is the most potent of the greenhouse gases in Earth's atmosphere, but its behaviour is fundamentally different from that of the other greenhouse gases. The primary role of water vapour is not as a direct agent of radiative forcing but rather as a climate feedback—that is, as a response within the climate system that influences the system's continued activity. This distinction arises from the fact that the amount of water vapour in the atmosphere cannot, in general, be directly modified by human behaviour but is instead set by air temperatures. The warmer the surface, the greater the evaporation rate of water from the surface. As a result, increased evaporation leads to a greater concentration of water vapour in the lower atmosphere capable of absorbing longwave radiation and emitting it downward.

Carbon Dioxide

Of the greenhouse gases, carbon dioxide (CO_2) is the most significant. Natural sources of atmospheric CO_2 include outgassing from volcanoes, the combustion and natural decay of organic matter, and respiration by aerobic (oxygen-using) organisms. These sources are balanced, on average, by a set of physical, chemical, or biological processes, called "sinks," that tend to remove CO_2 from the atmosphere. Significant natural sinks include terrestrial vegetation, which takes up CO_2 during the process of photosynthesis.

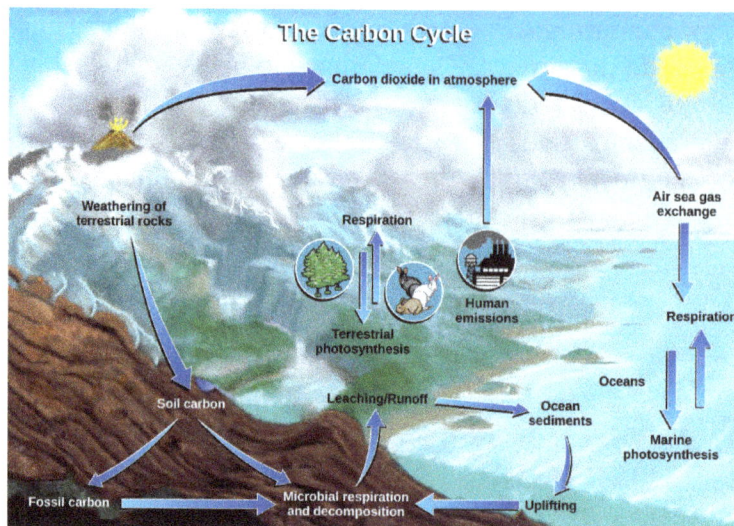

The carbon cycle.

Carbon is transported in various forms through the atmosphere, the hydrosphere, and geologic formations. One of the primary pathways for the exchange of carbon dioxide (CO_2) takes place between the atmosphere and the oceans; there a fraction of the CO_2 combines with water, forming carbonic acid (H_2CO_3) that subsequently loses hydrogen ions (H^+) to form bicarbonate (HCO_3^-) and carbonate (CO_3^{2-}) ions. Mollusk shells or mineral precipitates that form by the reaction of calcium or other metal ions with carbonate may become buried in geologic strata and eventually release CO_2 through volcanic outgassing. Carbon dioxide also exchanges through photosynthesis in plants and through respiration in animals. Dead and decaying organic matter may ferment and release CO_2 or methane (CH_4) or may be incorporated into sedimentary rock, where it is converted to fossil fuels. Burning of hydrocarbon fuels returns CO_2 and water (H_2O) to the atmosphere. The biological and anthropogenic pathways are much faster than the geochemical pathways and, consequently, have a greater impact on the composition and temperature of the atmosphere.

A number of oceanic processes also act as carbon sinks. One such process, called the "solubility pump," involves the descent of surface seawater containing dissolved CO_2. Another process, the "biological pump," involves the uptake of dissolved CO_2 by marine vegetation and phytoplankton (small free-floating photosynthetic organisms) living in the upper ocean or by other marine organisms that use CO_2 to build skeletons and other structures made of calcium carbonate ($CaCO_3$). As these organisms expire and fall to the ocean floor, the carbon they contain is transported downward and eventually buried at depth. A long-term balance between these natural sources and sinks leads to the background, or natural, level of CO_2 in the atmosphere.

In contrast, human activities increase atmospheric CO_2 levels primarily through the burning of fossil fuels—principally oil and coal and secondarily natural gas, for use in transportation, heating, and the generation of electrical power—and through the production of cement. Other anthropogenic sources include the burning of forests and the clearing of land. Anthropogenic emissions currently account for the annual release of about 7 gigatons (7 billion tons) of carbon into the atmosphere. Anthropogenic emissions are equal to approximately 3 percent of the total emissions of CO_2 by natural sources, and this amplified carbon load from human activities far exceeds the offsetting capacity of natural sinks (by perhaps as much as 2–3 gigatons per year).

Deforestation: Smoldering remains of a plot of deforested land in the Amazon Rainforest of Brazil. Annually, it is estimated that net global deforestation accounts for about two gigatons of carbon emissions to the atmosphere.

CO_2 consequently accumulated in the atmosphere at an average rate of 1.4 ppm per year between 1959 and 2006 and roughly 2.0 ppm per year between 2006 and 2018. Overall, this rate of accumulation has been linear (that is, uniform over time). However, certain current sinks, such as the oceans, could become sources in the future. This may lead to a situation in which the concentration of atmospheric CO_2 builds at an exponential rate (that is, its rate of increase is also increasing).

The natural background level of carbon dioxide varies on timescales of millions of years because of slow changes in outgassing through volcanic activity. For example, roughly 100 million years ago, during the Cretaceous Period (145 million to 66 million years ago), CO_2 concentrations appear to have been several times higher than they are today (perhaps close to 2,000 ppm). Over the past 700,000 years, CO_2 concentrations have varied over a far smaller range (between roughly 180 and 300 ppm) in association with the same Earth orbital effects linked to the coming and going of the Pleistocene ice ages. By the early 21st century, CO_2 levels had reached 384 ppm, which is approximately 37 percent above the natural background level of roughly 280 ppm that existed at the beginning of the Industrial Revolution. Atmospheric CO_2 levels continued to increase, and by 2018 they had reached 410 ppm. Such levels are believed to be the highest in at least 800,000 years according to ice core measurements and may be the highest in at least 5 million years according to other lines of evidence.

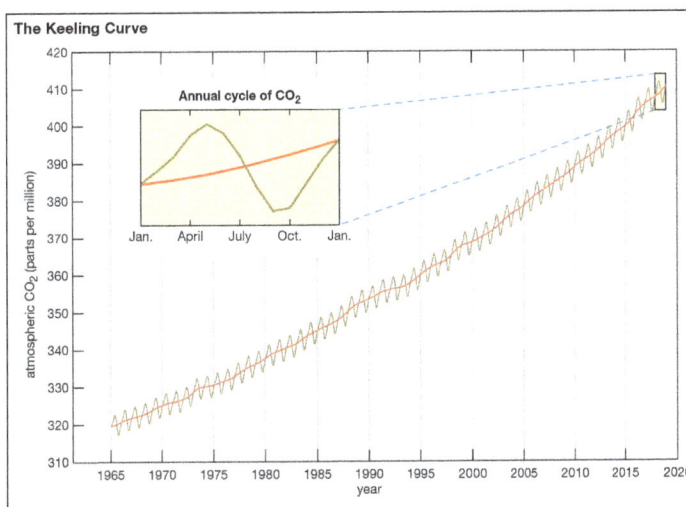

The Keeling Curve, named after American climate scientist Charles David Keeling, tracks changes in the concentration of carbon dioxide (CO_2) in Earth's atmosphere at a research station on Mauna Loa in Hawaii. Although these concentrations experience small seasonal fluctuations, the overall trend shows that CO_2 is increasing in the atmosphere.

Radiative forcing caused by carbon dioxide varies in an approximately logarithmic fashion with the concentration of that gas in the atmosphere. The logarithmic relationship occurs as the result of a saturation effect wherein it becomes increasingly difficult, as CO_2 concentrations increase, for additional CO_2 molecules to further influence the "infrared window" (a certain narrow band of wavelengths in the infrared region that is not absorbed by atmospheric gases). The logarithmic relationship predicts that the surface warming potential will rise by roughly the same amount for each doubling of CO_2 concentration. At current rates of fossil fuel use, a doubling of CO_2 concentrations over preindustrial levels is expected to take place by the middle of the 21st century (when CO_2 concentrations are projected to reach 560 ppm). A doubling of CO_2 concentrations would represent an increase of roughly 4 watts per square metre of radiative forcing. Given typical estimates of "climate sensitivity" in the absence of any offsetting factors, this energy increase would lead to a warming of 2 to 5 °C (3.6 to 9 °F) over preindustrial times. The total radiative forcing by anthropogenic CO_2 emissions since the beginning of the industrial age is approximately 1.66 watts per square metre.

Methane

Methane (CH_4) is the second most important greenhouse gas. CH_4 is more potent than CO_2 because the radiative forcing produced per molecule is greater. In addition, the infrared window is less saturated in the range of wavelengths of radiation absorbed by CH_4, so more molecules may fill in the region. However, CH_4 exists in far lower concentrations than CO_2 in the atmosphere, and its concentrations by volume in the atmosphere are generally measured in parts per billion (ppb) rather than ppm. CH_4 also has a considerably shorter residence time in the atmosphere than CO_2 (the residence time for CH_4 is roughly 10 years, compared with hundreds of years for CO_2).

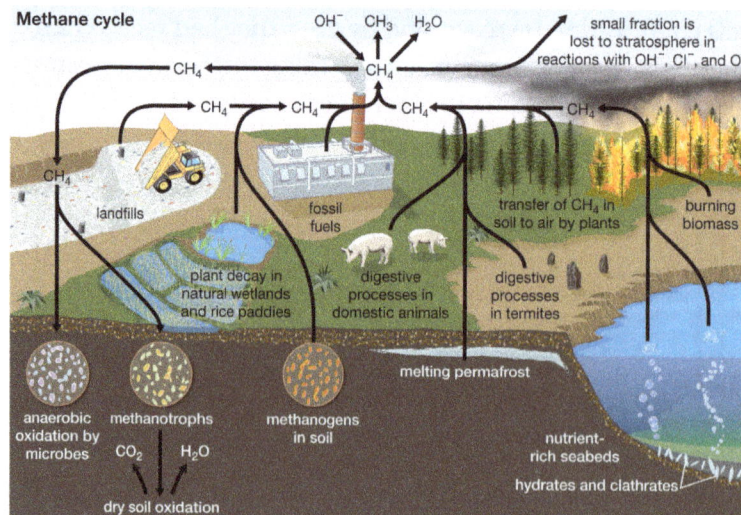

Methane cycle

Natural sources of methane include tropical and northern wetlands, methane-oxidizing bacteria that feed on organic material consumed by termites, volcanoes, seepage vents of the seafloor in regions rich with organic sediment, and methane hydrates trapped along the continental shelves of the oceans and in polar permafrost. The primary natural sink for methane is the atmosphere itself, as methane reacts readily with the hydroxyl radical ($\cdot$OH) within the troposphere to form CO_2 and water vapour (H_2O). When CH_4 reaches the stratosphere, it is destroyed. Another natural sink is soil, where methane is oxidized by bacteria.

As with CO_2, human activity is increasing the CH_4 concentration faster than it can be offset by natural sinks. Anthropogenic sources currently account for approximately 70 percent of total annual emissions, leading to substantial increases in concentration over time. The major anthropogenic sources of atmospheric CH_4 are rice cultivation, livestock farming, the burning of coal and natural gas, the combustion of biomass, and the decomposition of organic matter in landfills. Future trends are particularly difficult to anticipate. This is in part due to an incomplete understanding of the climate feedbacks associated with CH_4 emissions. In addition it is difficult to predict how, as human populations grow, possible changes in livestock raising, rice cultivation, and energy utilization will influence CH_4 emissions.

It is believed that a sudden increase in the concentration of methane in the atmosphere was responsible for a warming event that raised average global temperatures by 4–8 °C (7.2–14.4 °F) over a few thousand years during the so-called Paleocene-Eocene Thermal Maximum, or PETM. This episode took place roughly 55 million years ago, and the rise in CH_4 appears to have been related to a massive volcanic eruption that interacted with methane-containing flood deposits. As a result, large amounts of gaseous CH_4 were injected into the atmosphere. It is difficult to know precisely how high these concentrations were or how long they persisted. At very high concentrations, residence times of CH_4 in the atmosphere can become much greater than the nominal 10-year residence time that applies today. Nevertheless, it is likely that these concentrations reached several ppm during the PETM.

Methane concentrations have also varied over a smaller range (between roughly 350 and 800 ppb) in association with the Pleistocene ice age cycles. Preindustrial levels of CH_4 in the atmosphere were approximately 700 ppb, whereas levels exceeded 1,867 ppb in late 2018. (These concentrations are well above the natural levels observed for at least the past 650,000 years.) The net radiative forcing by anthropogenic CH_4 emissions is approximately 0.5 watt per square metre—or roughly one-third the radiative forcing of CO_2.

Surface-level Ozone and other Compounds

The next most significant greenhouse gas is surface, or low-level, ozone (O_3). Surface O_3 is a result of air pollution; it must be distinguished from naturally occurring stratospheric O_3, which has a very different role in the planetary radiation balance. The primary natural source of surface O_3 is the subsidence of stratospheric O_3 from the upper atmosphere. In contrast, the primary anthropogenic source of surface O_3 is photochemical reactions involving the atmospheric pollutant carbon monoxide (CO). The best estimates of the natural concentration of surface O_3 are 10 ppb, and the net radiative forcing due to anthropogenic emissions of surface O_3 is approximately 0.35 watt per square metre. Ozone concentrations can rise above unhealthy levels (that is, conditions where concentrations meet or exceed 70 ppb for eight hours or longer) in cities prone to photochemical smog.

Nitrous Oxides and Fluorinated Gases

Additional trace gases produced by industrial activity that have greenhouse properties include nitrous oxide (N_2O) and fluorinated gases (halocarbons), the latter including sulfur hexafluoride, hydrofluorocarbons (HFCs), and perfluorocarbons (PFCs). Nitrous oxide is responsible for 0.16 watt per square metre radiative forcing, while fluorinated gases are collectively responsible for 0.34 watt per square metre. Nitrous oxides have small background concentrations due to natural biological reactions in soil and water, whereas the fluorinated gases owe their existence almost entirely to industrial sources.

Aerosols

The production of aerosols represents an important anthropogenic radiative forcing of climate. Collectively, aerosols block—that is, reflect and absorb—a portion of incoming solar radiation, and this creates a negative radiative forcing. Aerosols are second only to greenhouse gases in relative importance in their impact on near-surface air temperatures. Unlike the decade-long residence times of the "well-mixed" greenhouse gases, such as CO_2 and CH_4, aerosols are readily flushed out of the atmosphere within days, either by rain or snow (wet deposition) or by settling out of the air (dry deposition). They must therefore be continually generated in order to produce a steady effect on radiative forcing. Aerosols have the ability to influence climate directly by absorbing or reflecting incoming solar radiation, but they can also produce indirect effects on climate by modifying cloud formation or cloud properties. Most aerosols serve as condensation nuclei (surfaces upon which water vapour can condense to form clouds); however, darker-coloured aerosols may hinder cloud formation by absorbing sunlight and heating up the surrounding air. Aerosols can be transported thousands of kilometres from their sources of origin by winds and upper-level circulation in the atmosphere.

Perhaps the most important type of anthropogenic aerosol in radiative forcing is sulfate aerosol. It is produced from sulfur dioxide (SO_2) emissions associated with the burning of coal and oil. Since the late 1980s, global emissions of SO_2 have decreased from about 151.5 million tonnes (167.0 million tons) to less than 100 million tonnes (110.2 million tons) of sulfur per year.

Nitrate aerosol is not as important as sulfate aerosol, but it has the potential to become a significant source of negative forcing. One major source of nitrate aerosol is smog (the combination of ozone with oxides of nitrogen in the lower atmosphere) released from the incomplete burning of fuel in internal-combustion engines. Another source is ammonia (NH_3), which is often used in fertilizers or released by the burning of plants and other organic materials. If greater amounts of atmospheric nitrogen are converted to ammonia and agricultural ammonia emissions continue to increase as projected, the influence of nitrate aerosols on radiative forcing is expected to grow.

Both sulfate and nitrate aerosols act primarily by reflecting incoming solar radiation, thereby reducing the amount of sunlight reaching the surface. Most aerosols, unlike greenhouse gases, impart a cooling rather than warming influence on Earth's surface. One prominent exception is carbonaceous aerosols such as carbon black or soot, which are produced by the burning of fossil fuels and biomass. Carbon black tends to absorb rather than reflect incident solar radiation, and so it has a warming impact on the lower atmosphere, where it resides. Because of its absorptive properties, carbon black is also capable of having an additional indirect effect on climate.

Through its deposition in snowfall, it can decrease the albedo of snow cover. This reduction in the amount of solar radiation reflected back to space by snow surfaces creates a minor positive radiative forcing.

Natural forms of aerosol include windblown mineral dust generated in arid and semiarid regions and sea salt produced by the action of waves breaking in the ocean. Changes to wind patterns as a result of climate modification could alter the emissions of these aerosols. The influence of climate change on regional patterns of aridity could shift both the sources and the destinations of dust clouds. In addition, since the concentration of sea salt aerosol, or sea aerosol, increases with the strength of the winds near the ocean surface, changes in wind speed due to global warming and climate change could influence the concentration of sea salt aerosol. For example, some studies suggest that climate change might lead to stronger winds over parts of the North Atlantic Ocean. Areas with stronger winds may experience an increase in the concentration of sea salt aerosol.

Other natural sources of aerosols include volcanic eruptions, which produce sulfate aerosol, and biogenic sources (e.g., phytoplankton), which produce dimethyl sulfide (DMS). Other important biogenic aerosols, such as terpenes, are produced naturally by certain kinds of trees or other plants. For example, the dense forests of the Blue Ridge Mountains of Virginia in the United States emit terpenes during the summer months, which in turn interact with the high humidity and warm temperatures to produce a natural photochemical smog. Anthropogenic pollutants such as nitrate and ozone, both of which serve as precursor molecules for the generation of biogenic aerosol, appear to have increased the rate of production of these aerosols severalfold. This process appears to be responsible for some of the increased aerosol pollution in regions undergoing rapid urbanization.

Human activity has greatly increased the amount of aerosol in the atmosphere compared with the background levels of preindustrial times. In contrast to the global effects of greenhouse gases, the impact of anthropogenic aerosols is confined primarily to the Northern Hemisphere, where most of the world's industrial activity occurs. The pattern of increases in anthropogenic aerosol over time is also somewhat different from that of greenhouse gases. During the middle of the 20th century, there was a substantial increase in aerosol emissions. This appears to have been at least partially responsible for a cessation of surface warming that took place in the Northern Hemisphere from the 1940s through the 1970s. Since that time, aerosol emissions have leveled off due to antipollution measures undertaken in the industrialized countries since the 1960s. Aerosol emissions may rise in the future, however, as a result of the rapid emergence of coal-fired electric power generation in China and India.

The total radiative forcing of all anthropogenic aerosols is approximately −1.2 watts per square metre. Of this total, −0.5 watt per square metre comes from direct effects (such as the reflection of solar energy back into space), and −0.7 watt per square metre comes from indirect effects (such as the influence of aerosols on cloud formation). This negative radiative forcing represents an offset of roughly 40 percent from the positive radiative forcing caused by human activity. However, the relative uncertainty in aerosol radiative forcing (approximately 90 percent) is much greater than that of greenhouse gases. In addition, future emissions of aerosols from human activities, and the influence of these emissions on future climate change, are not known with any certainty. Nevertheless, it can be said that, if concentrations of anthropogenic aerosols continue to decrease as they have since the 1970s, a significant offset to the effects of greenhouse gases will be reduced, opening future climate to further warming.

Land-use Change

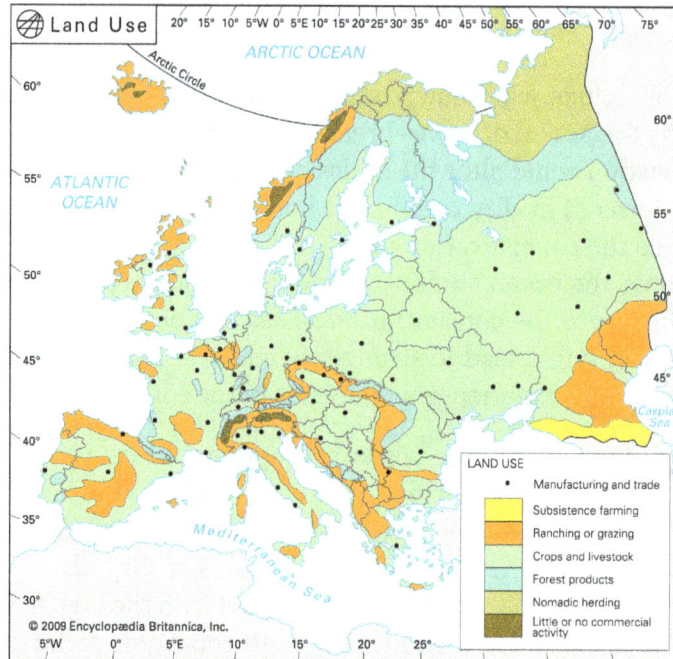

Land use in Europe.

There are a number of ways in which changes in land use can influence climate. The most direct influence is through the alteration of Earth's albedo, or surface reflectance. For example, the replacement of forest by cropland and pasture in the middle latitudes over the past several centuries has led to an increase in albedo, which in turn has led to greater reflection of incoming solar radiation in those regions. This replacement of forest by agriculture has been associated with a change in global average radiative forcing of approximately −0.2 watt per square metre since 1750. In Europe and other major agricultural regions, such land-use conversion began more than 1,000 years ago and has proceeded nearly to completion. For Europe, the negative radiative forcing due to land-use change has probably been substantial, perhaps approaching −5 watts per square metre. The influence of early land use on radiative forcing may help to explain a long period of cooling in Europe that followed a period of relatively mild conditions roughly 1,000 years ago. It is generally believed that the mild temperatures of this "medieval warm period," which was followed by a long period of cooling, rivaled those of 20th-century Europe.

Land-use changes can also influence climate through their influence on the exchange of heat between Earth's surface and the atmosphere. For example, vegetation helps to facilitate the evaporation of water into the atmosphere through evapotranspiration. In this process, plants take up liquid water from the soil through their root systems. Eventually this water is released through transpiration into the atmosphere, as water vapour through the stomata in leaves. While deforestation generally leads to surface cooling due to the albedo factor discussed above, the land surface may also be warmed as a result of the release of latent heat by the evapotranspiration process. The relative importance of these two factors, one exerting a cooling effect and the other a warming effect, varies by both season and region. While the albedo effect is likely to dominate in middle latitudes, especially during the period from autumn through spring, the evapotranspiration effect may dominate during the summer in the midlatitudes and year-round in the tropics. The latter case is particularly important in assessing the potential impacts of continued tropical deforestation.

The rate at which tropical regions are deforested is also relevant to the process of carbon sequestration, the long-term storage of carbon in underground cavities and biomass rather than in the atmosphere. By removing carbon from the atmosphere, carbon sequestration acts to mitigate global warming. Deforestation contributes to global warming, as fewer plants are available to take up carbon dioxide from the atmosphere. In addition, as fallen trees, shrubs, and other plants are burned or allowed to slowly decompose, they release as carbon dioxide the carbon they stored during their lifetimes. Furthermore, any land-use change that influences the amount, distribution, or type of vegetation in a region can affect the concentrations of biogenic aerosols, though the impact of such changes on climate is indirect and relatively minor.

Stratospheric Ozone Depletion

Since the 1970s the loss of ozone (O_3) from the stratosphere has led to a small amount of negative radiative forcing of the surface. This negative forcing represents a competition between two distinct effects caused by the fact that ozone absorbs solar radiation. In the first case, as ozone levels in the stratosphere are depleted, more solar radiation reaches Earth's surface. In the absence of any other influence, this rise in insolation would represent a positive radiative forcing of the surface. However, there is a second effect of ozone depletion that is related to its greenhouse properties. As the amount of ozone in the stratosphere is decreased, there is also less ozone to absorb longwave radiation emitted by Earth's surface. With less absorption of radiation by ozone, there is a corresponding decrease in the downward reemission of radiation. This second effect overwhelms the first and results in a modest negative radiative forcing of Earth's surface and a modest cooling of the lower stratosphere by approximately 0.5 °C (0.9 °F) per decade since the 1970s.

Natural Influences on Climate

There are a number of natural factors that influence Earth's climate. These factors include external influences such as explosive volcanic eruptions, natural variations in the output of the Sun, and slow changes in the configuration of Earth's orbit relative to the Sun. In addition, there are natural oscillations in Earth's climate that alter global patterns of wind circulation, precipitation, and surface temperatures. One such phenomenon is the El Niño/Southern Oscillation (ENSO), a coupled atmospheric and oceanic event that occurs in the Pacific Ocean every three to seven years. In addition, the Atlantic Multidecadal Oscillation (AMO) is a similar phenomenon that occurs over decades in the North Atlantic Ocean. Other types of oscillatory behaviour that produce dramatic shifts in climate may occur across timescales of centuries and millennia.

Volcanic Aerosols

Explosive volcanic eruptions have the potential to inject substantial amounts of sulfate aerosols into the lower stratosphere. In contrast to aerosol emissions in the lower troposphere, aerosols that enter the stratosphere may remain for several years before settling out, because of the relative absence of turbulent motions there. Consequently, aerosols from explosive volcanic eruptions have the potential to affect Earth's climate. Less-explosive eruptions, or eruptions that are less vertical in orientation, have a lower potential for substantial climate impact. Furthermore, because of large-scale circulation patterns within the stratosphere, aerosols injected within tropical regions tend to spread out over the globe, whereas aerosols injected within midlatitude and polar regions tend to remain confined to the middle and high latitudes of that hemisphere. Tropical eruptions, therefore, tend to have a greater climatic impact than eruptions occurring toward the poles. In 1991 the moderate eruption of Mount Pinatubo in the Philippines provided a peak forcing of approximately −4 watts per square metre and cooled the climate by about 0.5 °C (0.9 °F) over the following few years. By comparison, the 1815 Mount Tambora eruption in present-day Indonesia, typically implicated for the 1816 "year without a summer" in Europe and North America, is believed to have been associated with a radiative forcing of approximately −6 watts per square metre.

A column of gas and ash rising from Mount Pinatubo in the Philippines on June 12, 1991, just days before the volcano's climactic explosion on June 15.

While in the stratosphere, volcanic sulfate aerosol actually absorbs longwave radiation emitted by Earth's surface, and absorption in the stratosphere tends to result in a cooling of the troposphere below. This vertical pattern of temperature change in the atmosphere influences the behaviour of winds in the lower atmosphere, primarily in winter. Thus, while there is essentially a global cooling effect for the first few years following an explosive volcanic eruption, changes in the winter patterns of surface winds may actually lead to warmer winters in some areas, such as Europe. Some modern examples of major eruptions include Krakatoa (Indonesia) in 1883, El Chichón (Mexico) in 1982, and Mount Pinatubo in 1991. There is also evidence that volcanic eruptions may influence other climate phenomena such as ENSO.

Variations in Solar Output

Direct measurements of solar irradiance, or solar output, have been available from satellites only since the late 1970s. These measurements show a very small peak-to-peak variation in solar irradiance (roughly 0.1 percent of the 1,366 watts per square metre received at the top of the atmosphere, for approximately 1.4 watts per square metre). However, indirect measures of solar activity are available from historical sunspot measurements dating back through the early 17th century. Attempts have been made to reconstruct graphs of solar irradiance variations from historical sunspot data by calibrating them against the measurements from modern satellites. However, since the modern measurements span only a few of the most recent 11-year solar cycles, estimates of solar output variability on 100-year and longer timescales are poorly correlated. Different assumptions regarding the relationship between the amplitudes of 11-year solar cycles and long-period solar output changes can lead to considerable differences in the resulting solar reconstructions. These differences in turn lead to fairly large uncertainty in estimating positive forcing by changes in solar irradiance since 1750. (Estimates range from 0.06 to 0.3 watt per square metre.) Even more challenging, given the lack of any modern analog, is the estimation of solar irradiance during the so-called Maunder Minimum, a period lasting from the mid-17th century to the early 18th century when very few sunspots were observed. While it is likely that solar irradiance was reduced at this time, it is difficult to calculate by how much. However, additional proxies of solar output exist that match reasonably well with the sunspot-derived records following the Maunder Minimum; these may be used as crude estimates of the solar irradiance variations.

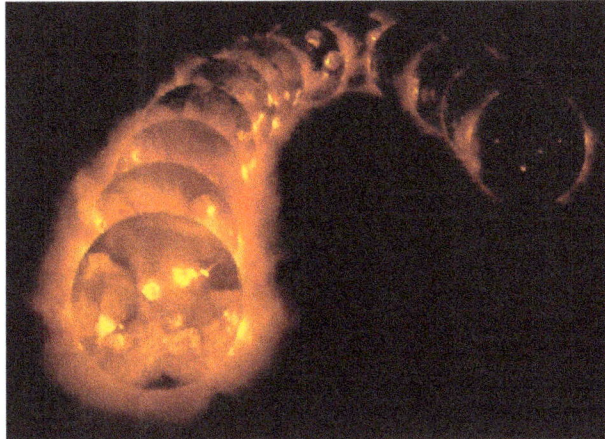

Twelve solar X-ray images obtained by Yohkoh between 1991 and 1995. The solar coronal brightness decreases by a factor of about 100 during a solar cycle as the Sun goes from an "active" state (left) to a less active state (right).

In theory it is possible to estimate solar irradiance even farther back in time, over at least the past millennium, by measuring levels of cosmogenic isotopes such as carbon-14 and beryllium-10. Cosmogenic isotopes are isotopes that are formed by interactions of cosmic rays with atomic nuclei in the atmosphere and that subsequently fall to Earth, where they can be measured in the annual layers found in ice cores. Since their production rate in the upper atmosphere is modulated by changes in solar activity, cosmogenic isotopes may be used as indirect indicators of solar irradiance. However, as with the sunspot data, there is still considerable uncertainty in the amplitude of past solar variability implied by these data.

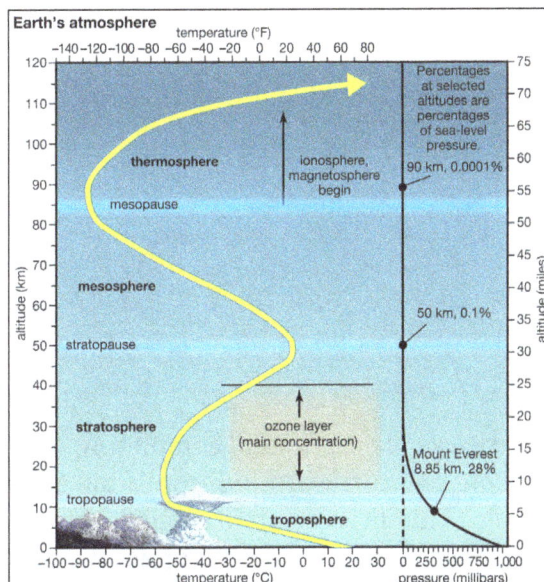

The layers of Earth's atmosphere. The yellow line shows the response of air temperature to increasing height.

Solar forcing also affects the photochemical reactions that manufacture ozone in the stratosphere. Through this modulation of stratospheric ozone concentrations, changes in solar irradiance (particularly in the ultraviolet portion of the electromagnetic spectrum) can modify how both shortwave and longwave radiation in the lower stratosphere are absorbed. As a result, the vertical temperature profile of the atmosphere can change, and this change can in turn influence phenomena such as the strength of the winter jet streams.

Variations in Earth's Orbit

On timescales of tens of millennia, the dominant radiative forcing of Earth's climate is associated with slow variations in the geometry of Earth's orbit about the Sun. These variations include the precession of the equinoxes (that is, changes in the timing of summer and winter), occurring on a roughly 26,000-year timescale; changes in the tilt angle of Earth's rotational axis relative to the plane of Earth's orbit around the Sun, occurring on a roughly 41,000-year timescale; and changes in the eccentricity (the departure from a perfect circle) of Earth's orbit around the Sun, occurring on a roughly 100,000-year timescale. Changes in eccentricity slightly influence the mean annual solar radiation at the top of Earth's atmosphere, but the primary influence of all the orbital variations listed above is on the seasonal and latitudinal distribution of incoming solar radiation over Earth's surface. The major ice ages of the Pleistocene Epoch were closely related to the influence of these variations on summer insolation at high northern latitudes. Orbital variations thus exerted a primary control on the extent of continental ice sheets. However, Earth's orbital changes are generally believed to have had little impact on climate over the past few millennia, and so they are not considered to be significant factors in present-day climate variability.

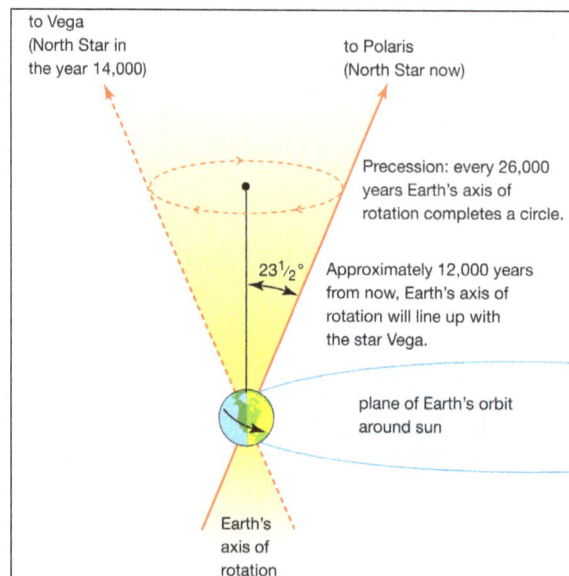

Earth's axis of rotation itself rotates, or precesses, completing one circle every 26,000 years. Consequently, Earth's North Pole points toward different stars (and sometimes toward empty space) as it travels in this circle. This precession is so slow that it is not noticeable in a person's lifetime, though astronomers must consider its effect when studying ancient sites such as Stonehenge.

Feedback Mechanisms and Climate Sensitivity

There are a number of feedback processes important to Earth's climate system and, in particular, its response to external radiative forcing. The most fundamental of these feedback mechanisms involves the loss of longwave radiation to space from the surface. Since this radiative loss increases with increasing surface temperatures according to the Stefan-Boltzmann law, it represents a stabilizing factor (that is, a negative feedback) with respect to near-surface air temperature.

Climate sensitivity can be defined as the amount of surface warming resulting from each additional watt per square metre of radiative forcing. Alternatively, it is sometimes defined as the warming that would result from a doubling of CO_2 concentrations and the associated addition of 4 watts per square metre of radiative forcing. In the absence of any additional feedbacks, climate sensitivity would be approximately 0.25 °C (0.45 °F) for each additional watt per square metre of radiative forcing. Stated alternatively, if the CO_2 concentration of the atmosphere present at the start of the industrial age (280 ppm) were doubled (to 560 ppm), the resulting additional 4 watts per square metre of radiative forcing would translate into a 1 °C (1.8 °F) increase in air temperature. However, there are additional feedbacks that exert a destabilizing, rather than stabilizing, influence, and these feedbacks tend to increase the sensitivity of climate to somewhere between 0.5 and 1.0 °C (0.9 and 1.8 °F) for each additional watt per square metre of radiative forcing.

Water Vapour Feedback

Unlike concentrations of other greenhouse gases, the concentration of water vapour in the atmosphere cannot freely vary. Instead, it is determined by the temperature of the lower atmosphere and surface through a physical relationship known as the Clausius-Clapeyron equation, named for 19th-century German physicist Rudolf Clausius and 19th-century French engineer Émile Clapeyron. Under the assumption that there is a liquid water surface in equilibrium with the atmosphere, this relationship indicates that an increase in the capacity of air to hold water vapour is a function of increasing temperature of that volume of air. This assumption is relatively good over the oceans, where water is plentiful, but not over the continents. For this reason the relative humidity (the percent of water vapour the air contains relative to its capacity) is approximately 100 percent over ocean regions and much lower over continental regions (approaching 0 percent in arid regions). Not surprisingly, the average relative humidity of Earth's lower atmosphere is similar to the fraction of Earth's surface covered by the oceans (that is, roughly 70 percent). This quantity is expected to remain approximately constant as Earth warms or cools. Slight changes to global relative humidity may result from human land-use modification, such as tropical deforestation and irrigation, which can affect the relative humidity over land areas up to regional scales.

The amount of water vapour in the atmosphere will rise as the temperature of the atmosphere rises. Since water vapour is a very potent greenhouse gas, even more potent than CO_2, the net greenhouse effect actually becomes stronger as the surface warms, which leads to even greater warming. This positive feedback is known as the "water vapour feedback." It is the primary reason that climate sensitivity is substantially greater than the previously stated theoretical value of 0.25 °C (0.45 °F) for each increase of 1 watt per square metre of radiative forcing.

Cloud Feedbacks

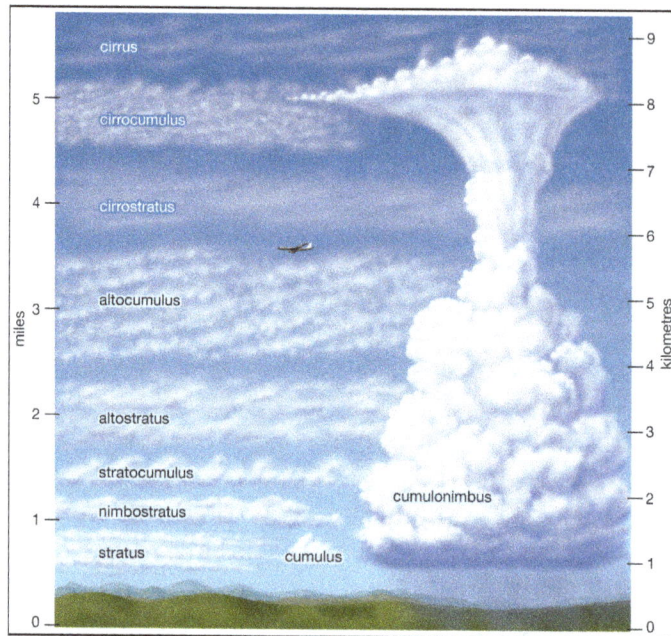

Different types of clouds form at different heights.

It is generally believed that as Earth's surface warms and the atmosphere's water vapour content increases, global cloud cover increases. However, the effects on near-surface air temperatures are complicated. In the case of low clouds, such as marine stratus clouds, the dominant radiative feature of the cloud is its albedo. Here any increase in low cloud cover acts in much the same way as an increase in surface ice cover: more incoming solar radiation is reflected and Earth's surface cools. On the other hand, high clouds, such as the towering cumulus clouds that extend up to the boundary between the troposphere and stratosphere, have a quite different impact on the surface radiation balance. The tops of cumulus clouds are considerably higher in the atmosphere and colder than their undersides. Cumulus cloud tops emit less longwave radiation out to space than the warmer cloud bottoms emit downward toward the surface. The end result of the formation of high cumulus clouds is greater warming at the surface.

The net feedback of clouds on rising surface temperatures is therefore somewhat uncertain. It represents a competition between the impacts of high and low clouds, and the balance is difficult to determine. Nonetheless, most estimates indicate that clouds on the whole represent a positive feedback and thus additional warming.

Ice Albedo Feedback

Another important positive climate feedback is the so-called ice albedo feedback. This feedback arises from the simple fact that ice is more reflective (that is, has a higher albedo) than land or water surfaces. Therefore, as global ice cover decreases, the reflectivity of Earth's surface decreases, more incoming solar radiation is absorbed by the surface, and the surface warms. This feedback is considerably more important when there is relatively extensive global ice cover, such as during the height of the last ice age, roughly 25,000 years ago. On a global scale the importance of ice albedo feedback decreases as Earth's surface warms and there is relatively less ice available to be melted.

Carbon Cycle Feedbacks

Another important set of climate feedbacks involves the global carbon cycle. In particular, the two main reservoirs of carbon in the climate system are the oceans and the terrestrial biosphere. These reservoirs have historically taken up large amounts of anthropogenic CO_2 emissions. Roughly 50–70 percent is removed by the oceans, whereas the remainder is taken up by the terrestrial biosphere. Global warming, however, could decrease the capacity of these reservoirs to sequester atmospheric CO_2. Reductions in the rate of carbon uptake by these reservoirs would increase the pace of CO_2 buildup in the atmosphere and represent yet another possible positive feedback to increased greenhouse gas concentrations.

In the world's oceans, this feedback effect might take several paths. First, as surface waters warm, they would hold less dissolved CO_2. Second, if more CO_2 were added to the atmosphere and taken up by the oceans, bicarbonate ions (HCO_3^-) would multiply and ocean acidity would increase. Since calcium carbonate ($CaCO_3$) is broken down by acidic solutions, rising acidity would threaten ocean-dwelling fauna that incorporate $CaCO_3$ into their skeletons or shells. As it becomes increasingly difficult for these organisms to absorb oceanic carbon, there would be a corresponding decrease in the efficiency of the biological pump that helps to maintain the oceans as a carbon sink. Third, rising surface temperatures might lead to a slowdown in the so-called thermohaline circulation, a global pattern of oceanic flow that partly drives the sinking of surface waters near the poles and is responsible for much of the burial of carbon in the deep ocean. A slowdown in this flow due to an influx of melting fresh water into what are normally saltwater conditions might also cause the solubility pump, which transfers CO_2 from shallow to deeper waters, to become less efficient. Indeed, it is predicted that if global warming continued to a certain point, the oceans would cease to be a net sink of CO_2 and would become a net source.

As large sections of tropical forest are lost because of the warming and drying of regions such as Amazonia, the overall capacity of plants to sequester atmospheric CO_2 would be reduced. As a result, the terrestrial biosphere, though currently a carbon sink, would become a carbon source. Ambient temperature is a significant factor affecting the pace of photosynthesis in plants, and many plant species that are well adapted to their local climatic conditions have maximized their photosynthetic rates. As temperatures increase and conditions begin to exceed the optimal temperature range for both photosynthesis and soil respiration, the rate of photosynthesis would decline. As dead plants decompose, microbial metabolic activity would increase and would eventually outpace photosynthesis.

Under sufficient global warming conditions, methane sinks in the oceans and terrestrial biosphere also might become methane sources. Annual emissions of methane by wetlands might either increase or decrease, depending on temperatures and input of nutrients, and it is possible that wetlands could switch from source to sink. There is also the potential for increased methane release as a result of the warming of Arctic permafrost (on land) and further methane release at the continental margins of the oceans (a few hundred metres below sea level). The current average atmospheric methane concentration of 1,750 ppb is equivalent to 3.5 gigatons (3.5 billion tons) of carbon. There are at least 400 gigatons of carbon equivalent stored in Arctic permafrost and as much as 10,000 gigatons (10 trillion tons) of carbon equivalent trapped on the continental margins of the oceans in a hydrated crystalline form known as clathrate. It is believed that some fraction of this trapped methane could become unstable with additional warming, although the amount and rate of potential emission remain highly uncertain.

Climate Research

Modern research into climatic variation and change is based on a variety of empirical and theoretical lines of inquiry. One line of inquiry is the analysis of data that record changes in atmosphere, oceans, and climate from roughly 1850 to the present. In a second line of inquiry, information describing paleoclimatic changes is gathered from "proxy," or indirect, sources such as ocean and lake sediments, pollen grains, corals, ice cores, and tree rings. Finally, a variety of theoretical models can be used to investigate the behaviour of Earth's climate under different conditions.

Modern Observations

Although a limited regional subset of land-based records is available from the 17th and 18th centuries, instrumental measurements of key climate variables have been collected systematically and at global scales since the mid-19th to early 20th century. These data include measurements of surface temperature on land and at sea, atmospheric pressure at sea level, precipitation over continents and oceans, sea ice extents, surface winds, humidity, and tides. Such records are the most reliable of all available climate data, since they are precisely dated and are based on well-understood instruments and physical principles. Corrections must be made for uncertainties in the data (for instance, gaps in the observational record, particularly during earlier years) and for systematic errors (such as an "urban heat island" bias in temperature measurements made on land).

Since the mid-20th century a variety of upper-air observations have become available (for example, of temperature, humidity, and winds), allowing climatic conditions to be characterized from the ground upward through the upper troposphere and lower stratosphere. Since the 1970s these data have been supplemented by polar-orbiting and geostationary satellites and by platforms in the oceans that gauge temperature, salinity, and other properties of seawater. Attempts have been made to fill the gaps in early measurements by using various statistical techniques and "backward prediction" models and by assimilating available observations into numerical weather prediction models. These techniques seek to estimate meteorological observations or atmospheric variables (such as relative humidity) that have been poorly measured in the past.

Modern measurements of greenhouse gas concentrations began with an investigation of atmospheric carbon dioxide (CO_2) concentrations by American climate scientist Charles Keeling at the summit of Mauna Loa in Hawaii in 1958. Keeling's findings indicated that CO_2 concentrations were steadily rising in association with the combustion of fossil fuels, and they also yielded the famous "Keeling curve," a graph in which the longer-term rising trend is superimposed on small oscillations related to seasonal variations in the uptake and release of CO_2 from photosynthesis and respiration in the terrestrial biosphere. Keeling's measurements at Mauna Loa apply primarily to the Northern Hemisphere.

Taking into account the uncertainties, the instrumental climate record indicates substantial trends since the end of the 19th century consistent with a warming Earth. These trends include a rise in global surface temperature of 0.9 °C (1.5 °F) between 1880 and 2012, an associated elevation of global sea level of 19–21 cm (7.5–8.3 inches) between 1901 and 2010, and a decrease in snow cover in the Northern Hemisphere of approximately 1.5 million square km (580,000 square miles). Records of average global temperatures kept by the World Meteorological Organization (WMO) indicate that the years 1998, 2005, and 2010 are statistically tied with one another as the warmest years since modern record keeping began in 1880; the WMO also noted that the decade 2001–10 was the warmest decade since 1880. Increases in global sea level are attributed to a combination of seawater expansion due to ocean heating and freshwater runoff caused by the melting of terrestrial ice. Reductions in snow cover are the result of warmer temperatures favouring a steadily shrinking winter season.

Climate data collected during the first two decades of the 21st century reveal that surface warming between 2005 and 2014 proceeded slightly more slowly than was expected from the effect of greenhouse gas increases alone. This fact was sometimes used to suggest that global warming had stopped or that it experienced a "hiatus" or "pause." In reality, this phenomenon appears to have been influenced by several factors, none of which, however, implies that global warming stopped during this period or that global warming would not continue in the future. One factor was the increased burial of heat beneath the ocean surface by strong trade winds, a process assisted by La Niña conditions. The effects of La Niña manifest in the form of cooling surface waters along the western coast of South America. As a result, warming at the ocean surface was reduced, but the accumulation of heat in other parts of the ocean occurred at an accelerated rate. Another factor cited by climatologists was a small but potentially important increase in aerosols from volcanic activity, which may have blocked a small portion of incoming solar radiation and which were accompanied by a small reduction in solar output during the period. These factors, along with natural decades-long oscillations in the climate system, may have masked a portion of the greenhouse warming. (However, climatologists point out that these natural climate cycles are expected to add to greenhouse warming in the future when the oscillations eventually reverse direction.) For these reasons many scientists believe that it is an error to call this slowdown in detectable surface warming a "hiatus" or a "pause."

Prehistorical Climate Records

In order to reconstruct climate changes that occurred prior to about the mid-19th century, it is necessary to use "proxy" measurements—that is, records of other natural phenomena that indirectly measure various climate conditions. Some proxies, such as most sediment cores and pollen records, glacial moraine evidence, and geothermal borehole temperature profiles, are coarsely resolved or dated and thus are only useful for describing climate changes on long timescales. Other proxies, such as growth rings from trees or oxygen isotopes from corals and ice cores, can provide a record of yearly or even seasonal climate changes.

The data from these proxies should be calibrated to known physical principles or related statistically to the records collected by modern instruments, such as satellites. Networks of proxy data can then be used to infer patterns of change in climate variables, such as the behaviour of surface temperature over time and geography. Yearly reconstructions of climate variables are possible over the past 1,000 to 2,000 years using annually dated proxy records, but reconstructions farther back in time are generally based on more coarsely resolved evidence such as ocean sediments and pollen records. For these, records of conditions can be reconstructed only on timescales of hundreds or thousands of years. In addition, since relatively few long-term proxy records are available for the Southern Hemisphere, most reconstructions focus on the Northern Hemisphere.

The various proxy-based reconstructions of the average surface temperature of the Northern Hemisphere differ in their details. These differences are the result of uncertainties implicit in the proxy data themselves and also of differences in the statistical methods used to relate the proxy data to surface temperature. Nevertheless, all studies as reviewed in the IPCC's Fourth Assessment Report (AR4), which was published in 2007, indicate that the average surface temperature since about 1950 is higher than at any time during the previous 1,000 years.

Theoretical Climate Models

Theoretical models of Earth's climate system can be used to investigate the response of climate to external radiative forcing as well as its own internal variability. Two or more models that focus on different physical processes may be coupled or linked together through a common feature, such as geographic location. Climate models vary considerably in their degree of complexity. The simplest models of energy balance describe Earth's surface as a globally uniform layer whose temperature is determined by a balance of incoming and outgoing shortwave and longwave radiation. These simple models may also consider the effects of greenhouse gases. At the other end of the spectrum are fully coupled, three-dimensional, global climate models. These are complex models that solve for radiative balance; for laws of motion governing the atmosphere, ocean, and ice; and for exchanges of energy and momentum within and between the different components of the climate. In some cases, theoretical climate models also include an interactive representation of Earth's biosphere and carbon cycle.

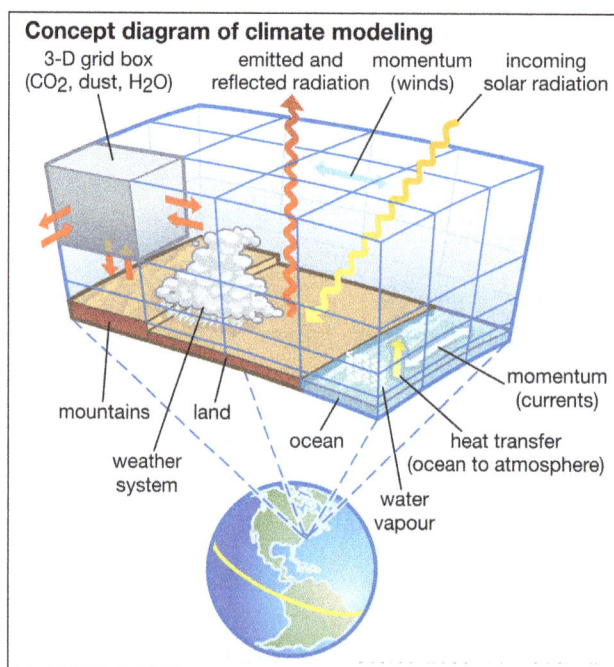

Concept diagram of climate modeling

3-D grid box (CO_2, dust, H_2O) · emitted and reflected radiation · momentum (winds) · incoming solar radiation · mountains · land · weather system · ocean · water vapour · heat transfer (ocean to atmosphere) · momentum (currents)

To understand and explain the complex behaviour of Earth's climate, modern climate models incorporate several variables that stand in for materials passing through Earth's atmosphere and oceans and the forces that affect them.

Even the most-detailed climate models cannot resolve all the processes that are important in the atmosphere and ocean. Most climate models are designed to gauge the behaviour of a number of physical variables over space and time, and they often artificially divide Earth's surface into a grid of many equal-sized "cells." Each cell may neatly correspond to some physical process (such as summer near-surface air temperature) or other variable (such as land-use type), and it may be assigned a relatively straightforward value. So-called "sub-grid-scale" processes, such as those of clouds, are too small to be captured by the relatively coarse spacing of the individual grid cells. Instead, such processes must be represented through a statistical process that relates the properties of the atmosphere and ocean. For example, the average fraction of cloud covers over a hypothetical "grid box" (that is, a representative volume of air or water in the model) can be estimated from the average relative humidity and the vertical temperature profile of the grid cell. Variations in the behaviour of different coupled climate models arise in large part from differences in the ways sub-grid-scale processes are mathematically expressed.

Despite these required simplifications, many theoretical climate models perform remarkably well when reproducing basic features of the atmosphere, such as the behaviour of midlatitude jet streams or Hadley cell circulation. The models also adequately reproduce important features of the oceans, such as the Gulf Stream. In addition, models are becoming better able to reproduce the main patterns of internal climate variability, such as those of El Niño/Southern Oscillation (ENSO). Consequently, periodically recurring events—such as ENSO and other interactions between the atmosphere and ocean currents—are being modeled with growing confidence.

Climate models have been tested in their ability to reproduce observed changes in response to radiative forcing. In 1988 a team at NASA's Goddard Institute for Space Studies in New York City used a fairly primitive climate model to predict warming patterns that might occur in response to three different scenarios of anthropogenic radiative forcing. Warming patterns were forecast for subsequent decades. Of the three scenarios, the middle one, which corresponds most closely to actual historical carbon emissions, comes closest to matching the observed warming of roughly 0.5 °C (0.9 °F) that has taken place since then. The NASA team also used a climate model to successfully predict that global mean surface temperatures would cool by about 0.5 °C for one to two years after the 1991 eruption of Mount Pinatubo in the Philippines.

More recently, so-called "detection and attribution" studies have been performed. These studies compare predicted changes in near-surface air temperature and other climate variables with patterns of change that have been observed for the past one to two centuries. The simulations have shown that the observed patterns of warming of Earth's surface and upper oceans, as well as changes in other climate phenomena such as prevailing winds and precipitation patterns, are consistent with the effects of an anthropogenic influence predicted by the climate models. In addition, climate model simulations have shown success in reproducing the magnitude and the spatial pattern of cooling in the Northern Hemisphere between roughly 1400 and 1850—during the Little Ice Age, which appears to have resulted from a combination of lowered solar output and heightened explosive volcanic activity.

Acid Rain

Acid rain, also called acid precipitation or acid deposition is the precipitation possessing a pH of about 5.2 or below primarily produced from the emission of sulfur dioxide (SO_2) and nitrogen oxides (NO_x; the combination of NO and NO_2) from human activities, mostly the combustion of fossil fuels. In acid-sensitive landscapes, acid deposition can reduce the pH of surface waters and lower biodiversity. It weakens trees and increases their susceptibility to damage from other stressors, such as drought, extreme cold, and pests. In acid-sensitive areas, acid rain also depletes soil of important plant nutrients and buffers, such as calcium and magnesium, and can release aluminum, bound to soil particles and rock, in its toxic dissolved form. Acid rain contributes to the corrosion of surfaces exposed to air pollution and is responsible for the deterioration of limestone and marble buildings and monuments.

The phrase acid rain was first used in 1852 by Scottish chemist Robert Angus Smith during his investigation of rainwater chemistry near industrial cities in England and Scotland. The phenomenon became an important part of his book *Air and Rain: The Beginnings of a Chemical Climatology* (1872). It was not until the late 1960s and early 1970s, however, that acid rain was recognized as a regional environmental issue affecting large areas of western Europe and eastern North America. Acid rain also occurs in Asia and parts of Africa, South America, and Australia. As a global environmental issue, it is frequently overshadowed by climate change. Although the problem of acid rain has been significantly reduced in some areas, it remains an important environmental issue within and downwind from major industrial and industrial agricultural regions worldwide.

Chemistry of Acid Deposition

Acid rain is a popular expression for the more scientific term acid deposition, which refers to the many ways in which acidity can move from the atmosphere to Earth's surface. Acid deposition includes acidic rain as well as other forms of acidic wet deposition—such as snow, sleet, hail, and fog (or cloud water). Acid deposition also includes the dry deposition of acidic particles and gases, which can affect landscapes during dry periods. Thus, acid deposition is capable of affecting landscapes and the living things that reside within them even when precipitation is not occurring.

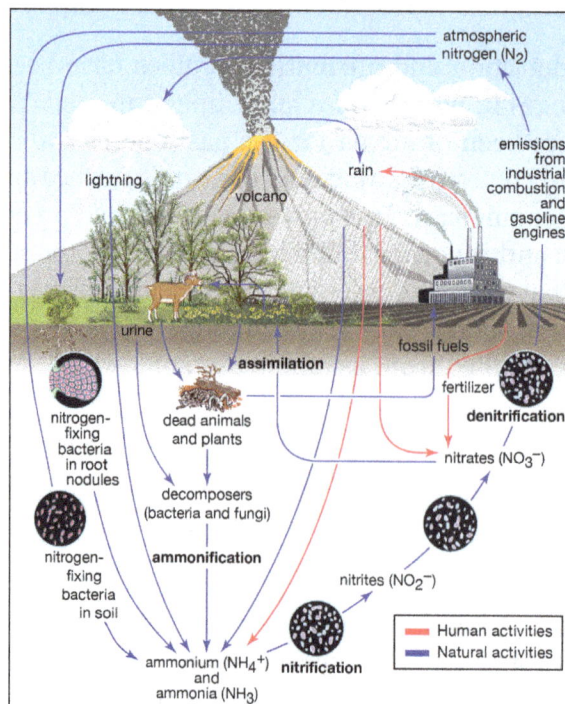

The nitrogen cycle.

Acidity is a measure of the concentration of hydrogen ions (H^+) in a solution. The pH scale measures whether a solution is acidic or basic. Substances are considered acidic below a pH of 7, and each unit of pH below 7 is 10 times more acidic, or has 10 times more H^+, than the unit above it. For example, rainwater with a pH of 5.0 has a concentration of 10 microequivalents of H^+ per litre, whereas rainwater with a pH of 4.0 has a concentration of 100 microequivalents of H^+ per litre.

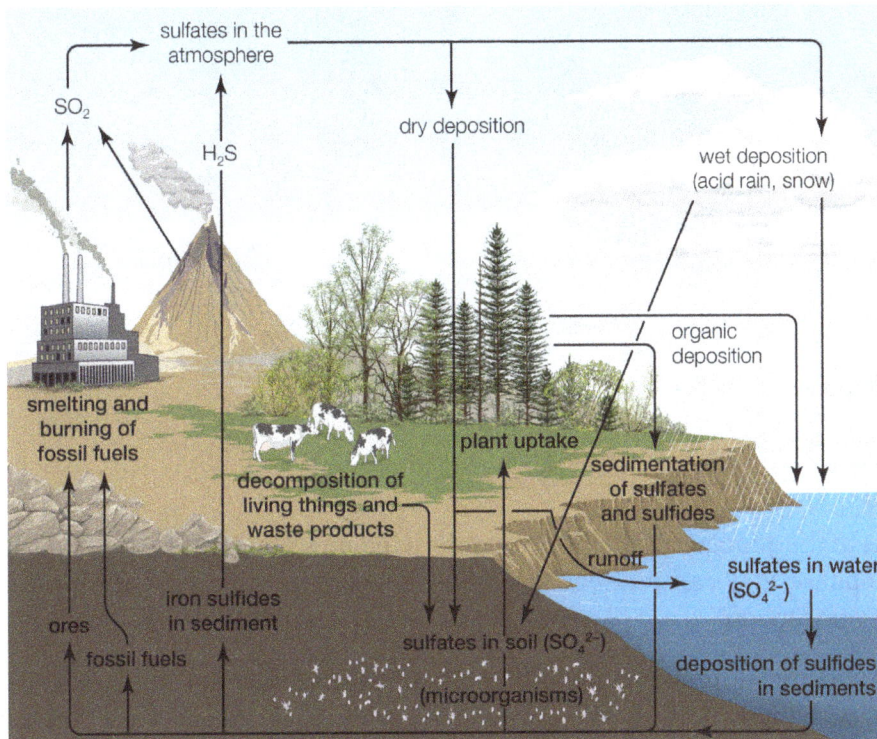

Sulfur cycle.

Major sulfur-producing sources include sedimentary rocks, which release hydrogen sulfide gas, and human sources, such as smelters and fossil-fuel combustion, both of which release sulfur dioxide into the atmosphere.

Normal rainwater is weakly acidic because of the absorption of carbon dioxide (CO_2) from the atmosphere—a process that produces carbonic acid—and from organic acids generated from biological activity. In addition, volcanic activity can produce sulfuric acid (H_2SO_4), nitric acid (HNO_3), and hydrochloric acid (HCl) depending on the emissions associated with specific volcanoes. Other natural sources of acidification include the production of nitrogen oxides from the conversion of atmospheric molecular nitrogen (N_2) by lightning and the conversion of organic nitrogen by wildfires. However, the geographic extent of any given natural source of acidification is small, and in most cases it lowers the pH of precipitation to no more than about 5.2.

Anthropogenic activities, particularly the burning of fossil fuels (coal, oil, natural gas) and the smelting of metal ores, are the major causes of acid deposition. In the United States, electric utilities produce nearly 70 percent of SO_2 and about 20 percent of NO_x emissions. Fossil fuels burned by vehicles account for nearly 60 percent of NO_x emissions in the United States. In the atmosphere, sulfuric and nitric acids are generated when SO_2 and NO_x, respectively, react with water. The simplest reactions are:

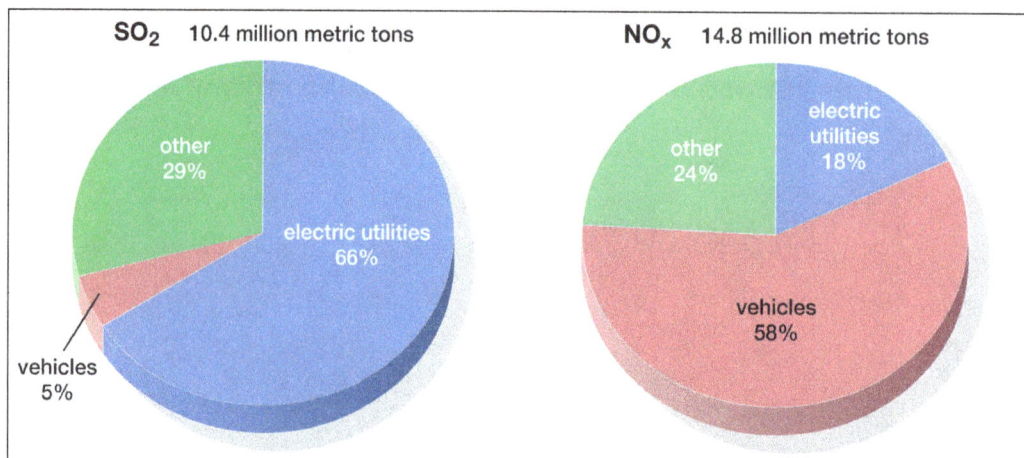

| SO₂ | 10.4 million metric tons |

other 29%

electric utilities 66%

vehicles 5%

NOₓ   14.8 million metric tons

electric utilities 18%

other 24%

vehicles 58%

U.S. SO_2 and NO_x emissions, 2008.

$$SO_2 + H_2O \rightarrow H_2SO_4 \longleftrightarrow H^+ + HSO_4 \longleftrightarrow 2H^+ + SO_4^2$$

$$NO_2 + H_2O \rightarrow HNO_3 \longleftrightarrow H^+ + NO_3$$

These reactions in the aqueous phase (for example, in cloud water) create wet deposition products. In the gaseous phase they can produce acidic dry deposition. Acid formation can also occur on particles in the atmosphere.

Where fossil fuel consumption is large and emission controls are not in place to reduce SO_2 and NO_x emissions, acid deposition will occur in areas downwind of emission sources, often hundreds to thousands of kilometres away. In such areas the pH of precipitation can average 4.0 to 4.5 annually, and the pH of individual rain events can sometimes drop below 3.0. In addition, cloud water and fog in polluted areas may be many times more acidic than rain falling over the same region.

Scientist testing the pH of polluted lake water containing melted acid snow.

Many air pollution and atmospheric deposition problems are intertwined with one another, and these problems are often derived from the same cause, namely the burning of fossil fuels. In addition to acid deposition, NO_x emissions along with hydrocarbon emissions are key ingredients in ground-level ozone (photochemical smog) formation, which is one of the most widespread forms of air pollution. The SO_2 and NO_x emissions can generate fine particulates, which are harmful to human respiratory systems. Coal combustion is the leading source of atmospheric mercury, which also enters ecosystems by wet and dry deposition. (A number of other heavy metals, such as lead and cadmium, and various particulates are also products of unregulated fossil fuel combustion.) Acid deposition of nitrogen derived from NO_x emissions creates additional environmental problems. For example, many lake, estuarine, and coastal marine systems receive too much nitrogen from atmospheric deposition and terrestrial runoff. This eutrophication (or over-enrichment) causes the overgrowth of plants and algae. When these organisms die and decompose, they deplete the dissolved oxygen supply necessary for most aquatic life in water bodies. Eutrophication is considered to be a major environmental problem in lake, coastal marine, and estuarine ecosystems worldwide.

Ecological Effects of Acid Deposition

Effects on lakes and Rivers

The regional effects of acid deposition were first noted in parts of western Europe and eastern North America in the late 1960s and early 1970s when changes in the chemistry of rivers and lakes, often in remote locations, were linked to declines in the health of aquatic organisms such as resident fish, crayfish, and clam populations. Increasing amounts of acid deposition in sensitive areas caused tens of thousands of lakes and streams in Europe and North America to become much more acidic than they had been in previous decades. Acid-sensitive areas are those that are predisposed to acidification because the region's soils have a low buffering capacity, or low acid-neutralizing capacity (ANC). In addition, acidification can release aluminum bound to soils, which in its dissolved form can be toxic to both plant and animal life. High concentrations of dissolved aluminum released from soils often enter streams and lakes. In conjunction with rising acidity in aquatic environments, aluminum can damage fish gills and thus impair respiration. In the Adirondack Mountain region of New York state, research has shown that the number of fish species drops from five in lakes with a pH of 6.0 to 7.0 to only one in lakes with a pH of 4.0 to 4.5. Other organisms are also negatively affected, so that acidified bodies of water lose plant and animal diversity overall. These effects can ripple throughout the food chain.

High acidity, especially from sulfur deposition, can accelerate the conversion of elemental mercury to its deadliest form: methyl mercury, a neurological toxin. This conversion most commonly occurs in wetlands and water-saturated soils where low-oxygen environments provide ideal conditions for the formation of methyl mercury by bacteria. Methyl mercury concentrates in organisms as it moves up the food chain, a phenomenon known as bioaccumulation. Small concentrations of methyl mercury present in phytoplankton and zooplankton accumulate in the fat cells of the animals that consume them. Since animals at higher tiers of the food chain must always consume large numbers of organisms from lower ones, the concentrations of methyl mercury in top predators, which often include humans, increase to levels where they could become harmful. The bioaccumulation of methyl mercury in the tissues of fishes is the leading reason for government health advisories that recommend reduced consumption of fish from fresh and marine waters.

In addition, aquatic acidification may be episodic, especially in colder climates. Sulfuric and nitric acid accumulating in a snowpack can leach out rapidly during the initial snowmelt and result in a pulse of acidic meltwater. Such pulses may be much more acidic than any individual snowfall event over the course of a winter, and these events can be deadly to acid-sensitive aquatic organisms throughout the food web.

Effects on Forested and Mountainous Regions

Spruce trees damaged by acid rain in Karkonosze National Park, Poland.

In the 1970s and '80s, forested areas in central Europe, southern Scandinavia, and eastern North America showed alarming signs of forest dieback and tree mortality. A 1993 survey in 27 European countries revealed air pollution damage or mortality in 23 percent of the 100,000 trees surveyed. It is likely that the dieback was the result of many factors, including acid deposition (e.g., soil acidification and loss of buffering capacity, mobilization of toxic aluminum, direct effects of acid on foliage), exposure to ground-level ozone, possible excess fertilization from the deposition of nitrogen compounds (such as nitrates, ammonium, and ammonia compounds), and general stress caused by a combination of these factors. Once a tree is in a weakened condition, it is more likely to succumb to other environmental stressors such as drought, insect infestation, and infection by pathogens. The areas of forest dieback were often found to be associated with regions with low buffering capacity where damage to aquatic ecosystems due to acid deposition was also occurring.

Acid deposition has been implicated in the alteration of soil chemistry and the decline of several tree species through both direct and indirect means. Poorly buffered soils are particularly susceptible to acidification because they lack significant amounts of base cations (positively charged ions), which neutralize acidity. Calcium, magnesium, sodium, and potassium, which are the base cations that account for most of the acid-neutralizing capacity of soils, are derived from the weathering of rocks and from wet and dry deposition. Some of these base cations (such as calcium and magnesium) are also secondary plant nutrients that are necessary for proper plant growth. The supply of these base cations declines as they neutralize the acids present in wet and dry deposition and are leached from the soils. Thus, a landscape formerly rich in base cations can become acid-sensitive when soil-formation processes are slow and base cations are not replaced through weathering or deposition processes.

Soil acidification can also occur where deposition of ammonia (NH_3) and ammonium (NH_4^+) is high. Ammonia and ammonium deposition leads to the production of H^+ (which results in acidification) when these chemicals are converted to nitrate (NO_3^-) by bacteria in a process called nitrification:

$$NH_3 + O_2 \rightarrow NO_2^- + 3H^+ + 2e^-$$

$$NO_2^- + H_2O \rightarrow NO_3^- + 2H^+ + 2e^-$$

The sources of NH_3 and NH_4^+ are largely agricultural activities, especially livestock (chickens, hogs, and cattle) production. Around 80 percent of NH_3 emissions in the United States and Europe come from the agricultural sector. The evaporation or volatilization of animal wastes releases NH_3 into the atmosphere. This process often results in the deposition of ammonia near the emission source. However, NH_3 can be converted to particulate ammonium that may be transported and deposited as wet and dry deposition hundreds of kilometres away from the emission source.

Besides negatively altering soil chemistry, acid deposition has been shown to affect some tree species directly. Red spruce (Picea rubens) trees found at higher elevations in the eastern United States are harmed by acids leaching calcium from the cell membranes in their needles, making the needles more susceptible to damage from freezing during winter. The damage is often greatest in mountainous regions, because these areas often receive more acid deposition than lower areas and the winter environment is more extreme. Mountainous regions are subjected to highly acidic cloud and fog water along with other environmental stresses. In addition, red spruce can be damaged by the increased concentration of toxic aluminum in the soil. These processes can reduce nutrient uptake by the tree roots. Sugar maple (Acer saccharum) populations are also declining in the northeastern United States and parts of eastern Canada. High soil aluminum and low soil calcium concentrations resulting from acid deposition have been implicated in this decline. Other trees in this region that are negatively affected by acidic deposition include aspen (Populus), birch (Betula), and ash (Fraxinus).

Some scientists argue that acid deposition may influence the geology of some regions. A 2018 study examining the 2009 Jiweishan landslide in southwest China proposed that acid rain may have weakened a layer of shale that separated the rock layers containing an aquifer above from the rock layers containing a mine below, which caused a large mass of rock to slip off the mountainside and kill 74 people.

Effects on Human-made Structures

Statue eroded by acid rain.

Acid deposition also affects human-made structures. The most notable effects occur on marble and limestone, which are common building materials found in many historic structures, monuments, and gravestones. Sulfur dioxide, an acid rain precursor, can react directly with limestone in the presence of water to form gypsum, which eventually flakes off or is dissolved by water. In addition, acid rain can dissolve limestone and marble through direct contact.

Wind Erosion

Wind erosion is a serious environmental problem and one of the most important natural causes of air pollution. There are places where this phenomenon is most likely to cause problems, especially in flat and bare areas or dry and sandy soils. However, to a greater or lesser extent, we all suffer from the air pollution this event provokes.

What is wind erosion? It is a natural process that moves soil from one location to another by wind power, often causing significant economic, health and environmental impact. At this point you might have thought about extreme cases, where a strong wind lifts a large volume of soil particulate matter into the air to create dust storms. However, light wind rolls soil particles along the surface, and this is the most common type of wind erosion.

At the end, it is wind that causes erosion, but the external facts, such as landscape or land condition, are the ones that determine the severity and the impact of this phenomenon. The following is a list of wind erosion examples:

- Rock formation

- Dunes

- Canyons

- Sand and dust storms.

After wind erosion, wind deposition occurs, and it is the geological process wherein soil particles or sediments are deposited and added to the mass of landforms.

Types of Wind Erosion

Suspension, Saltation and Surface Creep

There are 2 types to divide wind erosion, and one of them is through suspension, saltation and surface creep:

- Suspension: It is when particles are lifted into the wind, and once in the atmosphere, these dust and dirt particles can be transported very high through long distances, creating harmful environments for those who breath them.

- Saltation: It is when particles are lifted into the air, but this time they are drift horizontally. When these strike the ground again the velocity determines if they rebound back into the air or knock other particles into the air.

- Surface Creep: In this process, the particles are rolled across the surface because wind is not too strong or the particles are too heavy to be lifted.

Deflation and Abrasion

Deflation and abrasion are another way to categorize the types of wind erosion. Deflation occurs when wind moves particles that are loose and abrasion is when an area is eroded directly by airborne particles. In other words, deflation and abrasion indicate what agent is causing the erosion.

Causes of Wind Erosion

Wind erosion occurs when something causes a reduction to the ground cover below 50% or/and removes trees and scrub that act as windbreaks. Some example are land clearing, overgrazing by livestock or cropping.

However, as the name suggests, wind is the principal cause of erosion. It can happen anywhere and anytime the wind blows and it is stronger where the soil or sand is not compacted or is of a finely granulated nature.

What can Prevent Wind Erosion?

Wind erosion prevention is topic which governments globally should focus more, as it has a huge impact on land production. The loss of nutrients affects directly the ability of the soil to properly produce drops, and soil production is one of the main elements for human race to survive. The following facts are the five main ways to prevent or control wind erosion:

Surface/Crop Residues

The surface form and crop residues can help prevent wind erosion. If placed at the right angle, which is perpendicular to the existing wind, it protects the removal of soil particles and maintain the nutrients of the soil.

After the harvest, when the soil is highly exposed to wind erosion, it is recommended taking any harvest residues and spread them throw the soil, so these residues act as a protection layer for the soil particles and its nutrients.

Permanent Vegetation Cover

A permanent vegetation cover is not only for wind erosion protection, but also for the conservation of water and air resources. This vegetation cover includes growing grass, shrubs, trees, vegetables or legumes.

Surface Roughening

In large areas or areas where a permanent vegetation cover isn't enough to protect the soil from wind erosion, three extra surface roughening methods come up: soil crusts, crosswind ridge, and clod-forming tillage.

Reshaping the Land

Giving the land the ideal shape to protect it from wind erosion is key, especially on agricultural activities. It may not be available for everyone since it is a bit expensive in some cases, but it is a very effective way to lessen the potential for erosion.

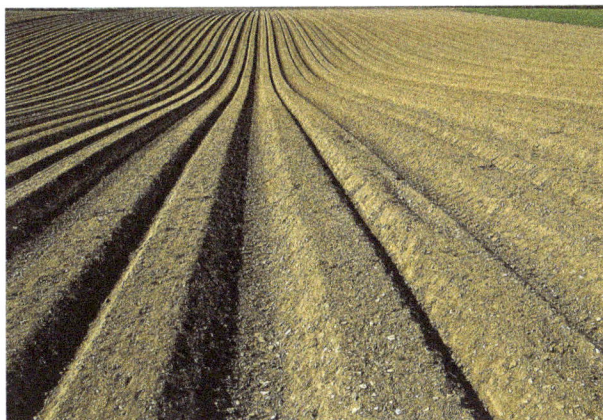

Irrigation

It is one of the best ways to lessen the erosion of soil since the wind force finds it difficult to carry the soil particles on a wet surface. However, too much water on a soil affects negatively the soil and its nutrients, and some very hot and dry areas cannot apply this technique.

Wind Erosion as an Air Pollution Source

Apart from the economic costs, airborne particulate matter and dust are harmful to humans when inhaled. It is highly recommended having a dust mask if you live in an area where dust storms are common.

Airborne dust is directly related to the probability of asthma and other health problems.

References

- What-air-pollution-does-to-your-body, entry: huffingtonpost.in, Retrieved 13 July, 2019

- Air-pollution-and-domestic-animals, air-pollution-new-developments, books: intechopen.com, Retrieved 24 March , 2019

- Environment-animals-plants-ecosystems, effects, air-pollution: airgo2.com, Retrieved 21 July, 2019

- Ozone-depletion, science: britannica.com, Retrieved 7 May, 2019

- Visibility, nightskies, subjects: nps.gov, Retrieved 13 June, 2019

- Global-warming, science: britannica.com, Retrieved 21 March, 2019

- Acid-rain, science: britannica.com, Retrieved 12 April, 2019

- Wind-erosion, natural, causes, air-pollution: airgo2.com, Retrieved 23 June, 2019

Chapter 5

Air Pollution: Modeling and Measurement

Air pollution modeling refers to the usage of mathematical theories in order to understand and predict the behavior or pollutants in the air. Measurement of air pollution includes air quality monitoring and maintaining the air quality index. The topics elaborated in this chapter will help in gaining a better perspective about the diverse aspects of air pollution modeling and measurement.

Air Pollution Modeling

Air pollution modeling is a numerical tool used to describe the causal relationship between emissions, meteorology, atmospheric concentrations, deposition, and other factors. Air pollution measurements give important, quantitative information about ambient concentrations and deposition, but they can only describe air quality at specific locations and times, without giving clear guidance on the identification of the causes of the air quality problem. Air pollution modeling, instead, can give a more complete deterministic description of the air quality problem, including an analysis of factors and causes (emission sources, meteorological processes, and physical and chemical changes), and some guidance on the implementation of mitigation measures.

Air pollution models play an important role in science, because of their capability to assess the relative importance of the relevant processes. Air pollution models are the only method that quantifies the deterministic relationship between emissions and concentrations/depositions, including the consequences of past and future scenarios and the determination of the effectiveness of abatement strategies. This makes air pollution models indispensable in regulatory, research, and forensic applications.

The concentrations of substances in the atmosphere are determined by: 1) transport, 2) diffusion, 3) chemical transformation, and 4) ground deposition. Transport phenomena, characterized by the mean velocity of the fluid, have been measured and studied for centuries. For example, the average wind has been studies by man for sailing purposes. The study of diffusion (turbulent motion) is more recent.

Modeling of Point Sources

One of the first challenges in the history of air pollution modeling was the understanding of the diffusion properties of plumes emitted from large industrial stacks. For this purpose, a very successful, yet simple model was developed – the Gaussian Plume Model. This model was applied for the main purpose of calculating the maximum ground level impact of plumes and the distance of maximum impact from the source.

The Gaussian plume model is illustrated in the figure below. The model was formulated by determining experimentally the horizontal and vertical spread of the plume, measured by the standard deviation of the plume's spatial concentration distribution.

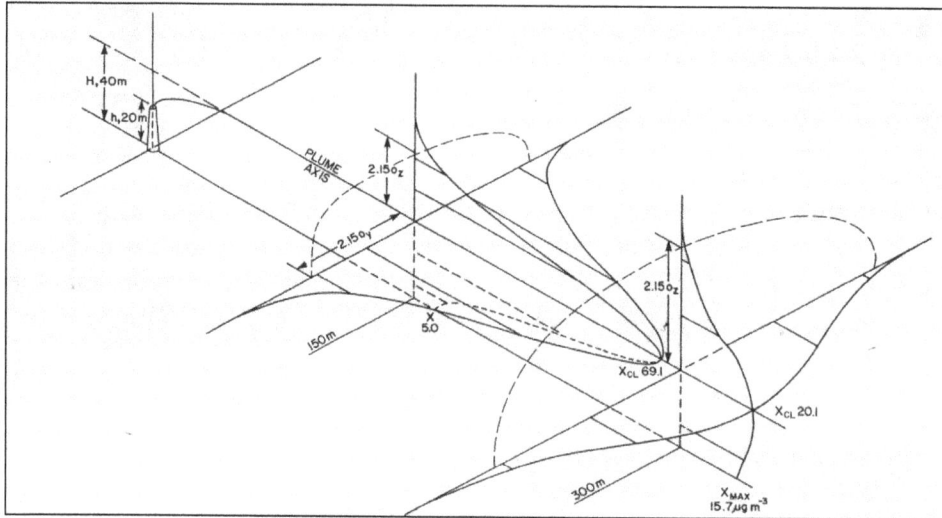

Two cross sections through a Gaussian plume (total mass under curves conserved).

Experiments provided the geometrical description of the plume by plotting the standard deviation of its concentration distribution, in both the vertical and horizontal direction, as a function of the atmospheric stability and downwind distance from the source. The plotting is presented in the figure below.

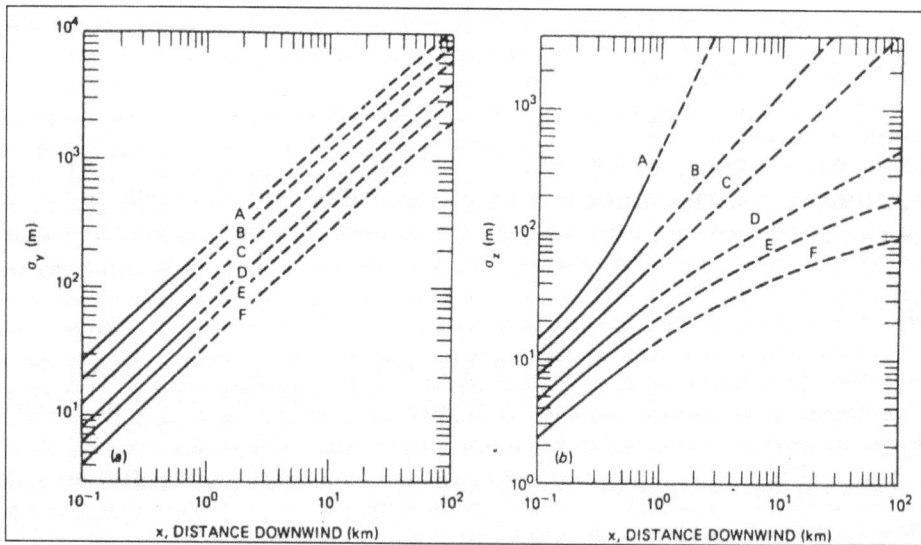

Pasquill-gifford σ_y (left) and σ_z (right).

Atmospheric stability is a parameter that characterizes the turbulent status of the atmosphere. This parameter ranges from "very stable", class F, to "neutral", class D, up to "very unstable", class A.

The experimental sigma values discussed above are, in their functions with distance from the source, in reasonable agreement with the Taylor-theory. The differences are caused by the fact that the Taylor-theory holds for homogeneous turbulence, which is not the exact case in the atmosphere.

In the 1960s, the studies concerning dispersion from a point source continued and were broadening in scope. Major studies were performed by Högstrom, Turner, Briggs (the developer of the well-known plume-rise formulas), Moore, and Klug. The use and application of the Gaussian plume model spread over the whole globe, and became a standard technique in every industrial country to calculate the stack height required for permits, for example Beryland who published a standard work in Russian. The Gaussian plume model concept was soon applied also to line and area-sources. Gradually, the importance of the mixing height was realized and its major influence on the magnitude of ground level concentrations. To include the effects of the mixing height, multiple reflections terms were added to the Gaussian Plume model.

Air Pollution Modeling at Urban and Larger Scales

Shortly after 1970, scientists began to realize that air pollution was not only a local phenomenon. It became clear - firstly in Europe - that the SO_2 and NO_x emissions from tall stacks could lead to acidification at large distances from the sources. It also became clear - firstly in the US - that ozone was a problem in urbanized and industrialized areas. And so it was obvious that these situations could not be tackled by simple Gaussian-plume type modeling.

Two different modeling approaches were followed, Lagrangian modeling and Eulerian modeling. In Lagrangian modeling, an air parcel (or "puff") is followed along a trajectory, and is assumed to keep its identity during its path. In Eulerian modeling, the area under investigation is divided into grid cells, both in vertical and horizontal directions.

Lagrangian modeling, directed at the description of long-range transport of sulfur, began with studies by Rohde, Eliassen and Fisher. The work by Eliassen was the start for the well-known EMEP-trajectory model which has been used over the years to calculate trans-boundary air pollution of acidifying species and later, photo-oxidants. Lagrangian modeling is often used to cover longer periods of time, up to years.

Eulerian modeling began with studies by Reynolds for ozone in urbanized areas, with Shir and Shieh for SO_2 in urban areas, and Egan and Carmichael for regional scale sulfur. From the modeling studies by Reynolds on the Los Angeles basin, the well-known Urban Airshed Model-UAM originated for photochemical simulations. Eulerian modeling, in these years, was used only for specific episodes of a few days.

So in general, Lagrangian modeling was mostly performed in Europe, over large distances and longer time-periods, and focused primarily on SO_2. Eulerian grid modeling was predominantly applied in the US, over urban areas and restricted to episodic conditions, and focused primarily on O_3. Also hybrid approaches were studied, as well as particle-in-cell methods. Early papers on both Eulerian and Lagrangian modeling are by Friedlander and Seinfeld, Eschenroeder and Martinez and Liu and Seinfeld.

A comprehensive overview of long-range transport modeling in the seventies was presented by Johnson.

The next, obvious step in scale is global modeling of earth's troposphere. The first global models were 2-D models, in which the global troposphere was averaged in the longitudinal direction. The first, 3-D global models were developed by Peters.

It can be stated that, since approximately 1980, the basic modeling concepts and tools were available to the scientific community. Developments after 1980 concerned the fine-tuning of these basic concepts.

Examples of Dispersion Models

We present below some discussion on specific air computer models that are particularly important and are used by a large community of scientists.

The US-EPA today recommends the following two computer packages for simulation of non-reactive chemicals (e.g., SO_2):

AERMOD

AERMOD is a steady-state Gaussian plume model. It uses a single wind field to transport emitted species. The wind field is derived from surface, upper-air, and onsite meteorological observations. AERMOD also combines geophysical data such as terrain elevations and land use with the meteorological data to derive boundary layer parameters such as Monin-Obukhov length, mixing height, stability class, turbulence, etc.

AERMOD is today replacing the ISC models for most regulatory applications in the US.

CALPUFF

CALPUFF is a non-steady state Lagrangian puff dispersion model. The advantage of this model over a Gaussian-based model is that is can realistically simulate the transport of substances in calm, stagnant conditions, complex terrain, and coastal regions with sea/land breezes.

CALPUFF is particularly recommended for long-range simulations (e.g., more than 50 miles) and studies involving the assessment of the visual impact of plumes.

With the development of the VISTAS Version 6 model, CALPUFF can use sub-hourly meteorological data and run with sub-hourly time steps. This version of CALPUFF is appropriate for both long-range and short-range simulations.

Photochemical Modeling

Photochemical air quality models have become widely recognized and routinely utilized tools for regulatory analysis and attainment demonstrations by assessing the effectiveness of control strategies. These photochemical models are large-scale air quality models that simulate the changes of pollutant concentrations in the atmosphere using a set of mathematical equations characterizing the chemical and physical processes in the atmosphere. These models are applied at multiple spatial scales from local, regional, national, and global.

Some examples of photochemical models are the following:

CMAQ

The primary goals for the Models-3/Community Multiscale Air Quality (CMAQ) modeling system are to improve: 1) the environmental management community's ability to evaluate the impact of air quality management practices for multiple pollutants at multiple scales and 2) the scientist's ability to better probe, understand, and simulate chemical and physical interactions in the atmosphere.

CAMX

The Comprehensive Air quality Model with extensions is a publicly available open-source computer modeling system for the integrated assessment of gaseous and particulate air pollution. Built on today's understanding that air quality issues are complex, interrelated, and reach beyond the urban scale, CAMx is designed to:

- Simulate air quality over many geographic scales.

- Treat a wide variety of inert and chemically active pollutants:

    ◦ Ozone.

    ◦ Inorganic and organic PM2.5/PM10.

    ◦ Mercury and toxics.

- Provide source-receptor, sensitivity, and process analyses.

- Be computationally efficient and easy to use.

The U.S. EPA has approved the use of CAMx for numerous ozone and PM State Implementation Plans throughout the U.S, and has used this model to evaluate regional mitigation strategies.

UAM

The Urban Airshed Model (UAM) modeling system, developed and maintained by Systems Applications International (SAI), is the most widely used photochemical air quality model in the world today. Since SAI's pioneering efforts in photochemical air quality modeling in the early 1970s, the model has undergone nearly continuous cycles of application, performance evaluation, update, extension, and improvement. Other photochemical models have been developed during this long period, but no model today is more reliable or technically superior.

Other Models

Many additional models are available either for regulatory applications or for R&D studies.

Meteorological Models

CALMET

CALMET is a meteorological diagnostic model that combines data from surface stations, upper-air stations, over-water stations, precipitation stations, with geophysical data like land use, terrain elevations, albedo, etc., to produce a fully 3-dimensional diagnostic gridded wind field for the duration of the CALPUFF simulation. This wind field is then passed into CALPUFF and is used to transport the emitted substances.

CALMET can link to prognostic meteorological models (i.e., MM5, ETA, RUC2, RAMS) and use their data to produce the gridded wind field.

MM5

The PSU/NCAR mesoscale model (known as MM5) is a limited-area, nonhydrostatic, terrain-following sigma-coordinate model designed to simulate or predict mesoscale atmospheric circulation. The model is supported by several pre- and post-processing programs, which are referred to collectively as the MM5 modeling system. The MM5 modeling system software is mostly written in Fortran, and has been developed at Penn State and NCAR as a community mesoscale model with contributions from users worldwide.

The MM5 modeling system software is freely provided and supported by the Mesoscale Prediction Group in the Mesoscale and Microscale Meteorology Division, NCAR.

RAMS

RAMS, the Regional Atmospheric Modeling System, is a highly versatile numerical code developed by scientists at Colorado State University for simulating and forecasting meteorological phenomena, and for depicting the results. Its major components are:

1. An atmospheric model which performs the actual simulations.

2. A data analysis package which prepares initial data for the atmospheric model from observed meteorological data.

3. A post-processing model visualization and analysis package, which interfaces atmospheric, model output with a variety of visualization software utilities.

Plume Rise Modules

Most air pollution models include a computational module for computing plume rise, i.e., the initial behavior of a hot plume injected vertically into a horizontal wind flow. In particular, AERMOD includes PRIME (Plume Rise Model Enhancements).

PRIME is an algorithm for simulating plume rise effects, including downwash as the plume travels over buildings.

Particle Models

Particle models are based on Lagrangian methods for simulating atmospheric diffusion. In these models, plumes are represented by thousands (even hundreds of thousands) of "fictitious" particles, which often move with semi-random trajectories in order to recreate the random components of atmospheric turbulence. These high-resolution models are particularly useful for simulating short-term releases from sources with highly variable emission rates in complex dispersion scenarios. Particle models are capable of simulating very short-term concentrations (e.g., 1-minute averages).

Examples are listed below:

- Kinematic Simulation Particle (KSP) Model in CALPUFF.

- Montecarlo.

Deposition Modules

Many air pollution models include a computational module for computing the fraction of the plume deposited at the ground as a consequence of dry and wet deposition phenomena.

Odor Modeling

The mechanisms of dispersion of odorous chemicals (e.g., mercaptans) in the atmosphere are the same as the dispersion of other pollutants. However, when multiple pollutants are emitted, masking and enhancing effects may occur. In this case, the relationship between concentrations of individual chemicals and odor is not well defined and odor must be characterized in terms of an odor detection threshold value for the entire mixture of odorous chemicals in the air. This is why, in odor modeling applications, it is often preferred to express the emission in "odor units".

Odor models must include algorithms to simulate instantaneous or semi-instantaneous concentrations, since odors are instantaneous human sensations.

Statistical Models

Statistical models are techniques based essentially on statistical data analysis of measured ambient concentrations. These models are not deterministic, in the sense that they do not establish nor simulate a cause-effect, physical relationship between emissions and ambient concentrations. Two main types of statistical models exist:

Air Quality Forecast and Alarm Systems

Statistical techniques (e.g., time series analysis, spectrazl analysis, Kalman filters) have been used to forecast air pollution trends a few hours in advance for the purpose of alerting the population or, for example, blocking automobile traffic.

Receptor Modeling

Receptor models are mathematical or statistical procedures for identifying and quantifying the sources of air pollutants at a receptor location. Unlike photochemical and dispersion air quality models, receptor models do not use pollutant emissions, meteorological data and chemical transformation mechanisms to estimate the contribution of sources to receptor concentrations. Instead, receptor models use the chemical and physical characteristics of gases and particles measured at source and receptor to both identify the presence of and to quantify source contributions to receptor concentrations.

Receptor models are mathematical or statistical procedures for identifying and quantifying the sources of air pollutants at a receptor location. Unlike photochemical and dispersion air quality models, receptor models do not use pollutant emissions, meteorological data and chemical transformation mechanisms to estimate the contribution of sources to receptor concentrations. Instead, receptor models use the chemical and physical characteristics of gases and particles measured at source and receptor to both identify the presence of and to quantify source contributions to receptor concentrations.

Modeling of Adverse Effects

Special models or mathematical techniques are available to calculate the adverse effects of air pollution. These models include:

- Health effects (e.g., cancer risk)

- Visibility impairment

- Global effects, such as climate change

- Damage to materials

- Ecological damages

Air Pollution Dispersion Model

Air quality models are used to predict ground level concentrations down point of sources. The object of a model is to relate mathematically the effects of source emissions on ground level concentrations, and to establish that permissible levels are, or are not, being exceeded. Models have been developed to meet these objectives for a variety of pollutants and time circumstances.

Models may be described according to the chemical reactions involved. So-called nonreactive models are applied to pollutants such as CO and SO_2 because of the simple manner in which their chemical reactions can be represented. Reactive models address complex multiple-species chemical mechanism common to atmospheric photochemistry and apply to pollutants such as NO, NO_2, and O_3.

Models can be described as simple or advanced based on the assumptions used and the degree of sophisticated with which the important variables are treated. Advanced models have been developed for such problems as photochemical pollution, dispersion in complex terrain, long-range transport, and point sources over flat terrain. The most widely used models for predicting the impact of relative unreactive gases, such as SO_2, released from smokestacks are based on Gaussian diffusion.

In Gaussian models, the spread of a plume in vertical horizontal directions is assumed to occur by simple diffusion along the direction of the mean wind. The maximum ground level concentration is calculated by means of the following equation:

$$C_x = \frac{Q}{\pi \sigma_y \sigma_{z^u}} e^{-1/2 \left[\frac{H}{\sigma_z} \right]^2} e^{-1/2 \left[\frac{y}{\sigma_y} \right]^2}$$

Key to stability classes					
Wind speed 10m		Day		Night	
(m/sec)	Incoming solar radiation			Thinly Overcast	
	Strong	Moderate	Slight	>4/8 Cloud	<3/8Cloud
<2	A	A-B	B	E	F
3-Feb	A-B	B	C	D	E
5-Mar	B	B-C	C	D	D
<6	C	D	D	D	D

Where

- C_x = ground level concentration at some distance x downwind (g/m3).

- Q = average emission rate (g/sec) u = mean wind speed (m/sec).

- H = effective stack height (m).

- σ_y = standard deviation of wind direction in the horizontal (m).

- σ_z = standard deviation of wind direction in the vertical (m).

- y = off-centerline distance (m).

- e= natural log equal to 2.71828.

The parameters σ_y and σ_z describe horizontal and vertical dispersion characteristics of a plume at various distances downwind of a source as function of different atmospheric stability conditions.

The effective stack height H is equal to the physical stack height (h) plus the height of the plume (plume rises, Δh) determined from where the plume bends over. Plume rises must be calculated from model equations before the effective stack height can be calculated.

For purposes of illustration, let us determine the ground level concentration (C_x) at some downwind distance (x). For the following conditions let us calculate the ground level concentrations at 10 km directly downwind.

A power plant burning 9 tons of 2.5% sulfur coal/hr emits SO_2 at a rate of 113 g/sec. The effective stack height is 100 m, and the wind speed is 3 m/sec. It is 1 hour before sunrise, and the sky is clear. Since the off centerline distance (Y) in this case is equal to O, the following equation reduces to:

$$C_x = \frac{Q}{\pi \sigma_y \sigma_z u} e^{-1/2 \left[\frac{H}{\sigma_z}\right]^2}$$

The atmospheric stability classes for the condition described is F. It represent a night time condition with <37.5% cloud cover. The horizontal dispersion coefficient σ_y for a downtime distance of 5 km for atmospheric stability class F is approximately 90 m; the vertical dispersion coefficient σ_z is approximately 20 m.

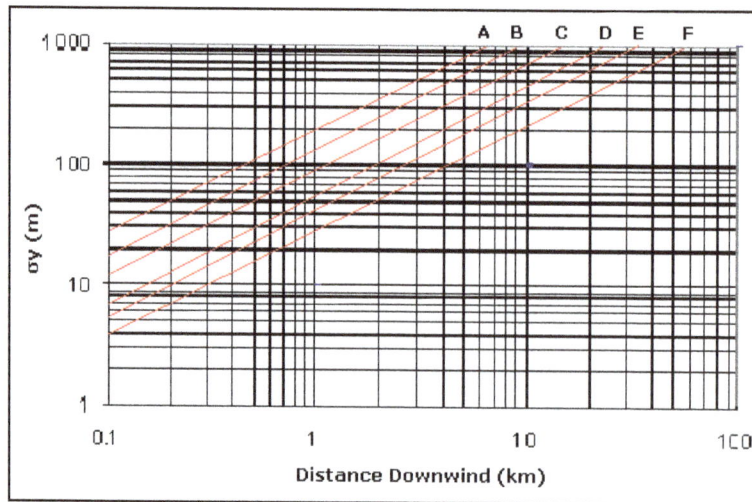

Horizontal Dispersion coefficent

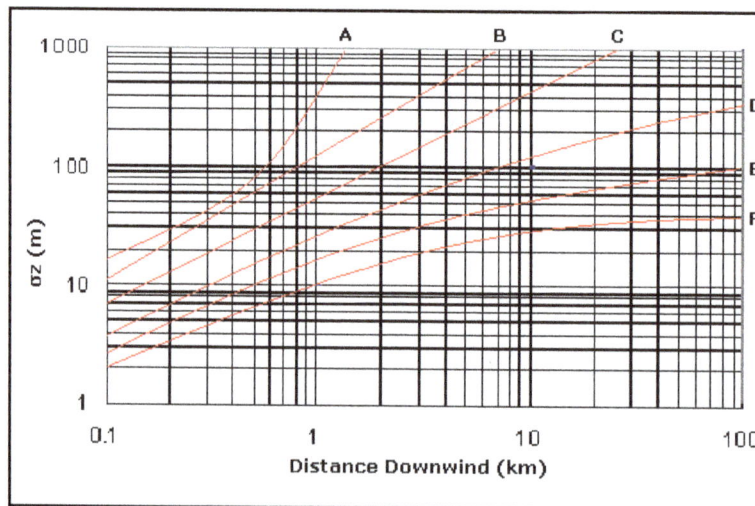

vertical dispersion coefficient

Therefore :

$$C_x = 44 \mu g / m^3$$

The ground level concentration of SO_2 from this source would be approximately 44 μg/m³ under the conditions given.

Although the use of air quality models is the subject of considerable controversy, there's a general agreement that there a few alternatives to the use of models, particulately to make decisions on an action which is know in advance to pose potential environmental problem. The debate arises as to which models should be used, and the interpretation of models results. The underlying question such in debates is how well, or how accurately, does the model predict concentrations under the specific circumstances, since model accuracy may vary from 30% to a factor of 2 or more? If a model is conservative, i.e., it over-predicts ground level concentrations, a source may be required to install costly control equipment unnecessarily. Less conservative models may under-predict concentrations and thus violations of air quality standards may occur. The uncertainty associated with input variables, such as wind data, and source emission data. Such data are usually estimated and not well documented.

Air Quality Index

Air quality index (AQI) is a numerical scale used for reporting day to day air quality with regard to human health and the environment. The daily results of the index are used to convey to the public an estimate of air pollution level. An increase in air quality index signifies increased air pollution and severe threats to human health. In most cases, AQI indicates how clear or polluted the air in our surrounding is, and the associated health risks it might present. The AQI centers on the health effects that may be experienced within a few days or hours after breathing polluted air.

AQI calculations focus on major air pollutants including: particulate matter, ground-level ozone, sulfur dioxide (SO_2), nitrogen dioxide (NO_2), and carbon monoxide (CO). Particulate matter and ozone pollutants pose the highest risks to human health and the environment. For each of these air pollutant categories, different countries have their own established air quality indices in relation to other nationally set air quality standards for public health protection.

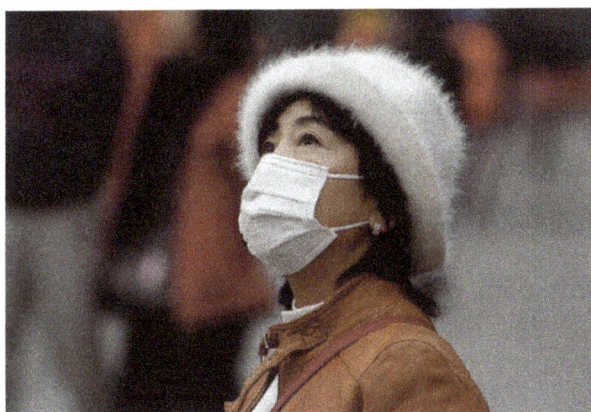

How does Air Quality Work?

On an hourly basis, the concentration of each pollutant in the air is measured and converted into a number running from zero upwards by using a standard index or scale. Calculated number for every pollutant is termed as a sub-index. The highest sub-index for any given hour is recorded as the AQI for that hour. In simple terms, AQI is like a yardstick that ranges from zero to five hundred (0-500). The index is a relative scale, meaning, the lower the index the better the quality of air and the lesser the health concern, and vice versa. The concentration of each pollutant varies, therefore, AQI values are grouped into ranges assigned to a standardized public health warnings and color code.

For instance, an AQI value of 0 to 50 means good air quality with limited possibility of affecting public and environmental health. On the other hand, an AQI value of 300 to 500 represents hazardous air quality with greater potential to affect public and environment health. The commonly accepted value is 100 as it corresponds to the generally approved air quality standards as set to safeguard public health. AQI levels below 100 are highly satisfactory while values beyond 100 are regarded to harm human health. As the AQI values get higher, it even poses more serious health concerns.

During days that the AQI is recorded to be elevated, the institution of public health might: offer advice to sensitive groups such as the children, those with respiratory problems, and the elderly to keep away from outdoor activities; take action to improve air quality by introducing measures for reducing emissions; or recommend the use of air pollution masks, especially in severe cases of air pollution.

Formula for Calculating Air Quality Index

Air quality index is calculated by a linear function by determining the concentration of the pollutant. The equation below is used to compute AQI.

$$I = \frac{I_{high} - I_{low}}{C_{high} - C_{low}}(C - C_{low}) + I_{low}$$

Where,

- I = the (Air Quality) index,

- C = the pollutant concentration,

- $C_{\{low\}}$ = the concentration breakpoint that is $\leq C$,

- $C_{\{high\}}$ = the concentration breakpoint that is $\geq C$,

- $I_{\{low\}}$ = the index breakpoint corresponding to $C_{\{low\}}$,

- $I_{\{high\}}$ = the index breakpoint corresponding to $C_{\{high\}}$.

Air Quality Index Categories

The AQI is divided in six categories and each category is meant to correspond to different health concern levels. Below is an explanation of the categories and their meanings.

- 0 – 50 indicates "Good" AQI. At this level the quality of air is deemed to be satisfactory, and air pollution poses little or no risk.

- 51 – 100 indicate "Moderate" AQI. This means acceptable Air quality. However, some pollutants might arouse modest health concern for a limited number of people. For instance, persons who are remarkably sensitive to ozone may experience respiratory symptoms.

- 101 – 150 indicate "Unhealthy for Sensitive Groups" AQI. This category may not be able to affect the general health of the public. However, children, older adults, and persons with lung disease are at a greater risk from ozone exposure. Older children, adults and people with lung and heart disease are at greater risk from exposure to particulate matter.

- 151 – 200 indicate "Unhealthy" AQI. In this category, every person might experience some adverse health effects. Sensitive group members may experience more serious effects.

People with heart or lung disease, older adults and children should cut back or reschedule strenuous activities.

- 201 – 300 indicate "Very Unhealthy" AQI. This would issue a health alert to mean that everybody may experience very serious health implications. People with heart or lung disease, older adults and children should significantly cut back or reschedule strenuous activities.

- Greater than 300 indicate "Hazardous" AQI. Air quality at this level is life-threatening and would issue warnings of emergency conditions for the entire population.

The table below provides the Air Quality Index (AQI) categories.

Air Quality Index (AQI) Values	Levels of Health Concern	Colors
When the AQI is in this range:	Air quality conditions are:	As symbolized by this color:
0-50	Good	Green
51-100	Moderate	Yellow
101-150	Unhealthy for Sensitive Groups	Orange
151 to 200	Unhealthy	Red
201 to 300	Very Unhealthy	Purple
301 to 500	Hazardous	Maroon

How to Avoid Exposure to Unhealthy Air?

The AQI is calculated for four major air pollutants regulated by the Clean Air Act: ground-level ozone, particle pollution, carbon monoxide, and sulfur dioxide. You need to take following simple steps to avoid exposure to unhealthy air:

- Prolonged Exertion: Prolonged exertion is any outdoor activity that you do intermittently for several hours and may cause you to breather slightly faster than normal. When air is unhealthy outside, you can reduce intake of unhealthy air by reducing how much time you spend on this type of activity.

- Heavy Exertion: Heavy exertion means intense outdoor activities that cause you to breathe hard. When air quality is bad outside, you can protect your health by reducing the amount of time you spend on this activity or by substituting it with less intense activity.

AQI Colors

Each AQI category is assigned a specific color to make it easier for people to understand the unhealthy levels of air pollution. For example, the color red means that conditions are "unhealthy for everyone."

Air Quality Index Levels of Health Concern	Numerical Value	Meaning
Good	0 to 50	Air quality is considered satisfactory, and air pollution poses little or no risk.
Moderate	51 to 100	Air quality is acceptable; however, for some pollutants there may be a moderate health concern for a very small number of people who are unusually sensitive to air pollution.
Unhealthy for Sensitive Groups	101 to 150	Members of sensitive groups may experience health effects. The general public is not likely to be affected.
Unhealthy	151 to 200	Everyone may begin to experience health effects; members of sensitive groups may experience more serious health effects.
Very Unhealthy	201 to 300	Health warnings of emergency conditions. The entire population is more likely to be affected.
Hazardous	301 to 500	Health alert: everyone may experience more serious health effects.

Air Quality Health Index

The Air Quality Health Index or "AQHI" is a scale designed to help you understand what the air quality around you means to your health.

It is a health protection tool that is designed to help you make decisions to protect your health by limiting short-term exposure to air pollution and adjusting your activity levels during increased levels of air pollution. It also provides advice on how you can improve the quality of the air you breathe.

This index pays particular attention to people who are sensitive to air pollution and provides them with advice on how to protect their health during air quality levels associated with low, moderate, high and very high health risks.

The AQHI communicates four primary things:

- Measures the air quality in relation to your health on a scale from 1 to 10. The higher the number, the greater the health risk associated with the air quality. When the amount of air pollution is very high, the number will be reported as 10+.

- Assigns a category that describes the level of health risk associated with the index reading (e.g. Low, Moderate, High, or Very High Health Risk).

- Provides health messages customized to each category for both the general population and the 'at risk' population.

- Shows current hourly AQHI readings and maximum forecast values for today, tonight and tomorrow.

The AQHI is designed to give you this information along with some suggestions on how you might adjust your activity levels depending on your individual health risk from air pollution.

How is the Air Quality Health Index Calculated?

The formula developed to calculate the Air Quality Health Index is based on research conducted by Health Canada using health and air quality data collected in major cities across Canada.

The Air Quality Health Index represents the relative risk of a mixture of common air pollutants which are known to harm human health. Three pollutants were chosen as indicators of the overall outdoor air mixture:

- Ground-level Ozone (O_3)

- Fine Particulate Matter ($PM_{2.5}$)

- Nitrogen Dioxide (NO_2)

What is the Scale for the Air Quality Health Index?

The Air Quality Health Index provides a number from 1 to 10+ to indicate the level of health risk associated with local air quality. Occasionally, when the amount of air pollution is abnormally high, the number may exceed 10. The higher the number, the greater the health risk and our need to take precautions.

The index describes the level of health risk associated with this number as 'low', 'moderate', 'high' or 'very high', and suggests steps we can take to reduce our exposure.

Air Quality Health Index Categories, Values and Associated Colours.

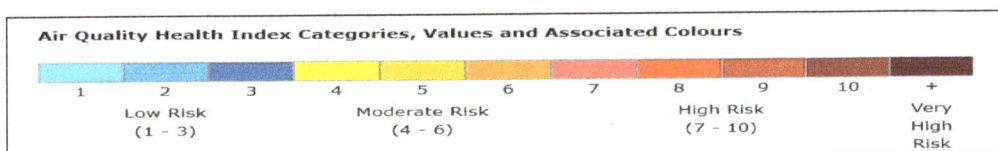

Air Quality Health Index Categories, Values and Associated Colours

| 1 | 2 | 3 | 4 | 5 | 6 | 7 | 8 | 9 | 10 | + |
| Low Risk (1 - 3) | | | Moderate Risk (4 - 6) | | | High Risk (7 - 10) | | | | Very High Risk |

- 1-3 Low health risk.

- 4-6 Moderate health risk.

- 7-10 High health risk.

- 10 + Very high health risk.

Air Quality Health Index Categories and Health Messages

The table below provides the health messages for each category of the Air Quality Health Index for the "at risk" population and the general population.

Health Risk	Air Quality Health Index	Health Messages	
		At Risk Population	General Population
Low	1 - 3	Enjoy your usual outdoor activities.	Ideal air quality for outdoor activities.
Moderate	4 - 6	Consider reducing or rescheduling strenuous activities outdoors if you are experiencing symptoms.	No need to modify your usual outdoor activities unless you experience symptoms such as coughing and throat irritation.
High	7 - 10	Reduce or reschedule strenuous activities outdoors. Children and the elderly should also take it easy.	Consider reducing or rescheduling strenuous activities outdoors if you experience symptoms such as coughing and throat irritation.
Very High	Above 10	Avoid strenuous activities outdoors. Children and the elderly should also avoid outdoor physical exertion.	Reduce or reschedule strenuous activities outdoors, especially if you experience symptoms such as coughing and throat irritation.

Air Quality Monitoring

The primary purpose of a systematic air quality monitoring network is to distinguish between areas where pollutant levels violate an ambient air quality standard and areas where they do not. As health-based ambient air quality standards are set at levels of pollutant concentrations that result in adverse impacts on human health, evidence of levels exceeding an ambient air quality standard in an area requires a public air quality agency to mitigate the corresponding pollutant. In other words, strategies, technologies, and regulations need to be developed to achieve the necessary reduction in pollution. The secondary purpose of a systematic monitoring network is to document the success of this sophisticated endeavor, either to record the rate of progress towards attaining the ambient air quality standard or to show that the standard has been achieved.

A systematic monitoring network in support of a successful implementation plan to reduce regional air pollution exposure must represent most of the population, and cover a diverse range of topography, meteorology, emissions, and air quality in the region. If possible, multiple pollutants and precursors need to be monitored at the same locations at the same time. For example, California requires annual evaluation of its air quality monitoring network, which consists of over 700 monitors at 250 sites, in order to document the current status and improvement of ambient air quality. The monitoring network represents various geographical areas, such as coastal areas, desert areas, interior valleys, mountain areas, and border areas, with more monitors in populated and polluted areas than in remote and clean areas. For individual monitors, spatial scales, whether they are microscale, middle scale, neighborhood scale, urban scale, or regional scale, are important considerations. In addition, diverse functions may be served by a monitor for the benefit of researchers, industries and media, as well as the general public.

Ambient Air Monitoring

Ambient air monitoring is the systematic, long-term assessment of pollutant levels by measuring the quantity and types of certain pollutants in the surrounding, outdoor air.

Ambient air monitoring is an integral part of an effective air quality management system. Reasons to collect such data include to:

- Assess the extent of pollution,

- Provide air pollution data to the general public in a timely manner,

- Support implementation of air quality goals or standards,

- Evaluate the effectiveness of emissions control strategies,

- Provide information on air quality trends,

- Provide data for the evaluation of air quality models and

- Support research (e.g., long-term studies of the health effects of air pollution).

There are different methods to measure any given pollutant. A developer of a monitoring strategy should examine the options to determine which methods are most appropriate, taking into account the main uses of the data, initial investment costs for equipment, operating costs, reliability of systems, and ease of operation.

The locations for monitoring stations depend on the purpose of the monitoring. Most air quality monitoring networks are designed to support human health objectives, and monitoring stations are established in population centers. They may be near busy roads, in city centers, or at locations of particular concern (e.g., a school, hospital, particular emissions sources). Monitoring stations also may be established to determine background pollution levels, away from urban areas and emissions sources.

Systems are needed to ensure that data are of acceptable quality, to record and store the data, and to analyze the data and present results.

Air Quality Monitor

An air quality monitor is a device that measures the level of common air pollutants. Monitors are available for both indoor and outdoor settings. Indoor air quality monitors are typically sensor based instruments. Some of them are able to measure ppb levels and come as either mixed gas or portable units. Sensor based instruments and air quality monitoring systems are used widely in outdoor ambient applications.

Industrial operators use air quality monitoring equipment to cost effectively monitor and manage emissions on their perimeter, which helps them improve relationships with regulators and communities. With air quality regulation shifting the burden from publicly funded monitoring to monitoring funded by industry, it has been increasingly important for businesses to acquire their own quality monitoring equipment.

Vehicular Air Pollution Information System

Vehicle exhaust emissions have always been considered the most important (and often the majority) source of air pollution in the cities. There are many ways to build an emissions inventory for this sector, depending on the level of information available.

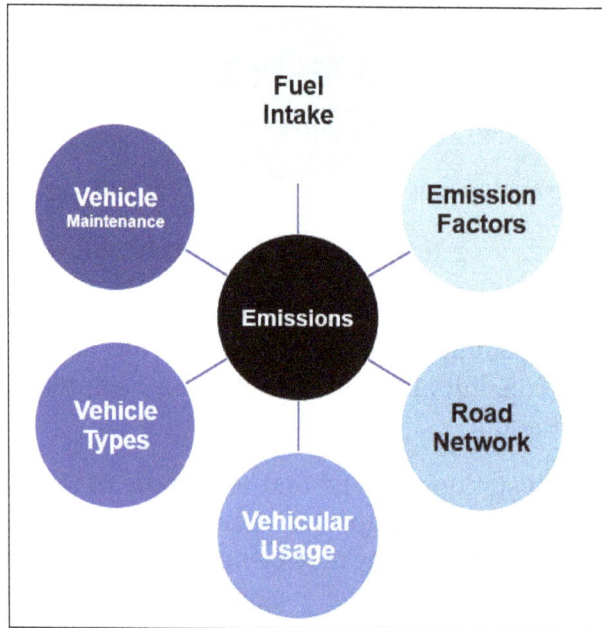

VAPIS tool, one of the integral parts of the SIM-air family of tools, is an user friendly spreadsheet based tool, designed to quickly analyze emission trends for a single vehicular category for a single pollutant. This tool allows the user to start with the basic information on vehicular numbers, growth rates, age splits, kilometers traveled per day, average retirement age, and emission factors, and establish the vehicular number and emission trends for a period of 30 years, by age group.

The fundamental equation utilized in VAPIS is the following:

References

- Air-quality-dispersion-model, air-purification: lenntech.com, Retrieved 2 June, 2019

- Air-quality-index: conserve-energy-future.com, Retrieved 17 March, 2019

- Aqhi-description, science: airqualityontario.com, Retrieved 13 February, 2019

- Air-quality-monitoring, earth-and-planetary-sciences, topics: sciencedirect.com, Retrieved 7 May, 2019

- Managing-air-quality-ambient-air-monitoring, air-quality-management-process: epa.gov, Retrieved 13 July, 2019

Chapter 6

Air Pollution Control

Air pollution control refers to the techniques which are used to remove the harmful substances which pollute the air. The chapter closely examines several focus areas related to air pollution control such as the equipment used for air pollution control and the diverse ways to stop air pollution.

Air pollution control are the techniques employed to reduce or eliminate the emission into the atmosphere of substances that can harm the environment or human health. The control of air pollution is one of the principal areas of pollution control, along with wastewater treatment, solid-waste management, and hazardous-waste management.

Air is considered to be polluted when it contains certain substances in concentrations high enough and for durations long enough to cause harm or undesirable effects. These include adverse effects on human health, property, and atmospheric visibility. The atmosphere is susceptible to pollution from natural sources as well as from human activities. Some natural phenomena, such as volcanic eruptions and forest fires, may have not only local and regional effects but also long-lasting global ones. Nevertheless, only pollution caused by human activities, such as industry and transportation, is subject to mitigation and control.

Most air contaminants originate from combustion processes. During the Middle Ages the burning of coal for fuel caused recurrent air pollution problems in London and other large European cities. Beginning in the 19th century, in the wake of the Industrial Revolution, increasing use of fossil fuels intensified the severity and frequency of air pollution episodes. The advent of mobile sources of air pollution—i.e., gasoline-powered highway vehicles—had a tremendous impact on air quality problems in cities. It was not until the middle of the 20th century, however, that meaningful and lasting attempts were made to regulate or limit emissions of air pollutants from stationary and mobile sources and to control air quality on both regional and local scales.

The primary focus of air pollution regulation in industrialized countries has been on protecting ambient, or outdoor, air quality. This involves the control of a small number of specific "criteria" pollutants known to contribute to urban smog and chronic public health problems. The criteria pollutants include fine particulates, carbon monoxide, sulfur dioxide, nitrogen dioxide, ozone, and lead. Since the end of the 20th century, there also has been recognition of the hazardous effects of trace amounts of many other air pollutants called "air toxics." Most air toxics are organic chemicals, comprising molecules that contain carbon, hydrogen, and other atoms. Specific emission regulations have been implemented against those pollutants. In addition, the long-term and far-reaching effects of the "greenhouse gases" on atmospheric chemistry and climate have been observed, and cooperative international efforts have been undertaken to control those pollutants. The greenhouse gases include carbon dioxide, chlorofluorocarbons (CFCs), methane, nitrous oxide, and ozone. In 2009 the U.S. Environmental Protection Agency ruled that greenhouse gases posed a threat to human health and could be subject to regulation as air pollutants.

The best way to protect air quality is to reduce the emission of pollutants by changing to cleaner fuels and processes. Pollutants not eliminated in this way must be collected or trapped by appropriate air-cleaning devices as they are generated and before they can escape into the atmosphere.

Control of Particulates

Airborne particles can be removed from a polluted airstream by a variety of physical processes. Common types of equipment for collecting fine particulates include cyclones, scrubbers, electrostatic precipitators, and baghouse filters. Once collected, particulates adhere to each other, forming agglomerates that can readily be removed from the equipment and disposed of, usually in a landfill.

Because each air pollution control project is unique, it is usually not possible to decide in advance what the best type of particle-collection device (or combination of devices) will be; control systems must be designed on a case-by-case basis. Important particulate characteristics that influence the selection of collection devices include corrosivity, reactivity, shape, density, and especially size and size distribution (the range of different particle sizes in the airstream). Other design factors include airstream characteristics (e.g., pressure, temperature, and viscosity), flow rate, removal efficiency requirements, and allowable resistance to airflow. In general, cyclone collectors are often used to control industrial dust emissions and as pre-cleaners for other kinds of collection devices. Wet scrubbers are usually applied in the control of flammable or explosive dusts or mists from such sources as industrial and chemical processing facilities and hazardous-waste incinerators; they can handle hot airstreams and sticky particles. Electrostatic precipitators and fabric-filter baghouses are often used at power plants.

Cyclones

A cyclone removes particulates by causing the dirty airstream to flow in a spiral path inside a cylindrical chamber. Dirty air enters the chamber from a tangential direction at the outer wall of the device, forming a vortex as it swirls within the chamber. The larger particulates, because of their greater inertia, move outward and are forced against the chamber wall. Slowed by friction with the wall surface, they then slide down the wall into a conical dust hopper at the bottom of the cyclone. The cleaned air swirls upward in a narrower spiral through an inner cylinder and emerges from an outlet at the top. Accumulated particulate dust is periodically removed from the hopper for disposal.

Cyclones are best at removing relatively coarse particulates. They can routinely achieve efficiencies of 90 percent for particles larger than about 20 micrometres (μm; 20 millionths of a metre). By themselves, however, cyclones are not sufficient to meet stringent air quality standards. They are typically used as pre-cleaners and are followed by more efficient air-cleaning equipment such as electrostatic precipitators and baghouses.

Electrostatic Precipitators

Electrostatic precipitation is a commonly used method for removing fine particulates from air-streams. In an electrostatic precipitator, particles suspended in the airstream are given an electric charge as they enter the unit and are then removed by the influence of an electric field. The precipitation unit comprises baffles for distributing airflow, discharge and collection electrodes, a dust clean-out system, and collection hoppers. A high voltage of direct current (DC), as much as 100,000 volts, is applied to the discharge electrodes to charge the particles, which then are attracted to oppositely charged collection electrodes, on which they become trapped.

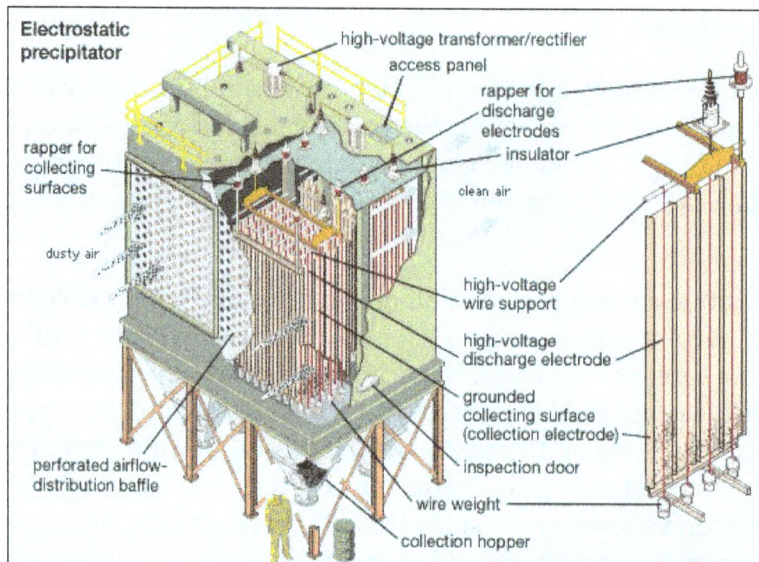

Electrostatic precipitator, a common particle-collection device at fossil-fuel power-generating stations.

In a typical unit the collection electrodes comprise a group of large rectangular metal plates suspended vertically and parallel to each other inside a boxlike structure. There are often hundreds of plates having a combined surface area of tens of thousands of square metres. Rows of discharge electrode wires hang between the collection plates. The wires are given a negative electric charge, whereas the plates are grounded and thus become positively charged.

Particles that stick to the collection plates are removed periodically when the plates are shaken, or "rapped." Rapping is a mechanical technique for separating the trapped particles from the plates, which typically become covered with a 6-mm (0.2-inch) layer of dust. Rappers are either of the impulse (single-blow) or vibrating type. The dislodged particles are collected in a hopper at the bottom of the unit and removed for disposal. An electrostatic precipitator can remove particulates as small as 1 μm with an efficiency exceeding 99 percent. The effectiveness of electrostatic precipitators in removing fly ash from the combustion gases of fossil-fuel furnaces accounts for their high frequency of use at power stations.

Baghouse Filters

One of the most efficient devices for removing suspended particulates is an assembly of fabric-filter bags, commonly called a baghouse. A typical baghouse comprises an array of long, narrow bags—each about 25 cm (10 inches) in diameter—that are suspended upside down in a large enclosure. Dust-laden air is blown upward through the bottom of the enclosure by fans. Particulates are trapped inside the filter bags, while the clean air passes through the fabric and exits at the top of the baghouse.

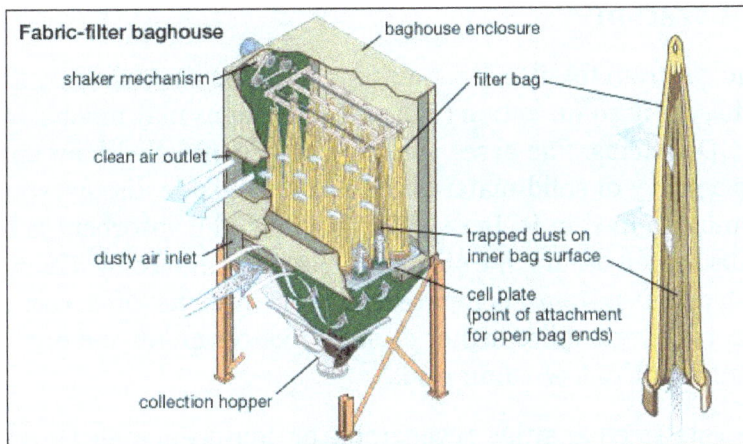

Baghouse employing an array of fabric bags for filtering the airstream.

A fabric-filter dust collector can remove very nearly 100 percent of particles as small as 1 μm and a significant fraction of particles as small as 0.01 μm. Fabric filters, however, offer relatively high resistance to airflow, which leads to substantial energy usage for the fan system. In addition, in order to prolong the useful life of the filter fabric, the air to be cleaned must be cooled (usually below 300 °C [570 °F]) before it is passed through the unit; cooling coils needed for this purpose add to the energy usage. (Certain filter fabrics—e.g., those made of ceramic or mineral materials—can operate at higher temperatures.)

Several compartments of filter bags are often used at a single baghouse installation. This arrangement allows individual compartments to be cleaned while others remain in service. The bags are cleaned by removing the excess layer of surface dust. This is done in several different ways: by mechanically shaking them; by temporarily reversing the flow of air and causing them to collapse; or by sending a short burst of air down through the bag, causing it to briefly expand. After the dust is removed from the filters, it falls into a hopper below and can be collected for disposal or further use. Care must be taken not to remove too much of the built-up surface dust, or "dust cake," when cleaning the filters. In most filter types the filter itself is only a substrate that allows for the formation of a layer of dust cake, which then captures the majority of the particulates. Filters with an applied membrane coating such as polytetrafluoroethylene (Teflon) do not require the use of dust cake to operate at their highest efficiency.

Control of Gases

Gaseous criteria pollutants, as well as volatile organic compounds (VOCs) and other gaseous air toxics, are controlled by means of three basic techniques: absorption, adsorption, and incineration (or combustion). These techniques can be employed singly or in combination. They are effective against the major greenhouse gases as well. In addition, a fourth technique, known as carbon sequestration, is in development as a means of controlling carbon dioxide levels.

Absorption

In the context of air pollution control, absorption involves the transfer of a gaseous pollutant from the air into a contacting liquid, such as water. The liquid must be able either to serve as a solvent for the pollutant or to capture it by means of a chemical reaction.

Flue Gas Desulfurization

Sulfur dioxide in flue gas from fossil-fuel power plants can be controlled by means of an absorption process called flue gas desulfurization (FGD). FGD systems may involve wet scrubbing or dry scrubbing. In wet FGD systems, flue gases are brought in contact with an absorbent, which can be either a liquid or a slurry of solid material. The sulfur dioxide dissolves in or reacts with the absorbent and becomes trapped in it. In dry FGD systems, the absorbent is dry pulverized lime or limestone; once absorption occurs, the solid particles are removed by means of baghouse filters (described above). Dry FGD systems, compared with wet systems, offer cost and energy savings and easier operation, but they require higher chemical consumption and are limited to flue gases derived from the combustion of low-sulfur coal.

FGD systems are also classified as either regenerable or nonregenerable (throwaway), depending on whether the sulfur that is removed from the flue gas is recovered or discarded. In the United States most systems in operation are nonregenerable because of their lower capital and operating costs. By contrast, in Japan regenerable systems are used extensively, and in Germany they are required by law. Nonregenerable FGD systems produce a sulfur-containing sludge residue that requires appropriate disposal. Regenerable FGD systems require additional steps to convert the sulfur dioxide into useful by-products like sulfuric acid.

Several FGD methods exist, differing mainly in the chemicals used in the process. FGD processes that employ either lime or limestone slurries as the reactants are widely applied. In the limestone scrubbing process, sulfur dioxide reacts with limestone (calcium carbonate) particles in the slurry, forming calcium sulfite and carbon dioxide. In the lime scrubbing process, sulfur dioxide reacts with slaked lime (calcium hydroxide), forming calcium sulfite and water. Depending on sulfur dioxide concentrations and oxidation conditions, the calcium sulfite can continue to react with water, forming calcium sulfate (gypsum). Neither calcium sulfite nor calcium sulfate is very soluble in water, nor both can be precipitated out as a slurry by gravity settling. The thick slurry, called FGD sludge, creates a significant disposal problem. Flue gas desulfurization helps to reduce ambient sulfur dioxide levels and mitigate the problem of acid rain. Nevertheless, in addition to its expense (which is passed on directly to the consumer as higher rates for electricity), millions of tons of FGD sludge are generated each year.

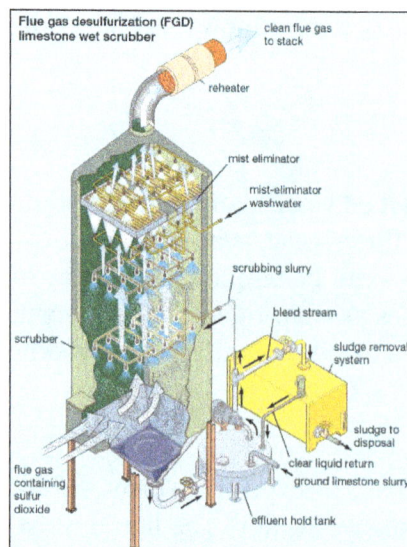

Wet scrubber using a limestone slurry to remove sulfur dioxide from flue gas.

Adsorption

Gas adsorption, as contrasted with absorption, is a surface phenomenon. The gas molecules are sorbed—attracted to and held—on the surface of a solid. Gas adsorption methods are used for odour control at various types of chemical-manufacturing and food-processing facilities, in the recovery of a number of volatile solvents (e.g., benzene), and in the control of VOCs at industrial facilities.

Activated carbon (heated charcoal) is one of the most common adsorbent materials. It is very porous and has an extremely high ratio of surface area to volume. Activated carbon is particularly useful as an adsorbent for cleaning airstreams that contain VOCs and for solvent recovery and odour control. A properly designed carbon adsorption unit can remove gas with an efficiency exceeding 95 percent.

Adsorption systems are configured either as stationary bed units or as moving bed units. In stationary bed adsorbers, the polluted airstream enters from the top, passes through a layer, or bed, of activated carbon, and exits at the bottom. In moving bed adsorbers, the activated carbon moves slowly down through channels by gravity as the air to be cleaned passes through in a cross-flow current.

Incineration

The process called incineration or combustion—chemically, rapid oxidation—can be used to convert VOCs and other gaseous hydrocarbon pollutants to carbon dioxide and water. Incineration of VOCs and hydrocarbon fumes usually is accomplished in a special incinerator called an afterburner. To achieve complete combustion, the afterburner must provide the proper amount of turbulence and burning time, and it must maintain a sufficiently high temperature. Sufficient turbulence, or mixing, is a key factor in combustion because it reduces the required burning time and temperature. A process called direct flame incineration can be used when the waste gas is itself a combustible mixture and does not need the addition of air or fuel.

An afterburner typically is made of a steel shell lined with refractory material such as firebrick. The refractory lining protects the shell and serves as a thermal insulator. Given enough time and high enough temperatures, gaseous organic pollutants can be almost completely oxidized, with incineration efficiency approaching 100 percent. Certain substances, such as platinum, can act in a manner that assists the combustion reaction. These substances, called catalysts, allow complete oxidation of the combustible gases at relatively low temperatures.

Afterburners are used to control odours, destroy toxic compounds, or reduce the amount of photochemically reactive substances released into the air. They are employed at a variety of industrial facilities where VOC vapours are emitted from combustion processes or solvent evaporation (e.g., petroleum refineries, paint-drying facilities, and paper mills).

Carbon Sequestration

The best way to reduce the levels of carbon dioxide in the air is to use energy more efficiently and to reduce the combustion of fossil fuels by using alternative energy sources (e.g., nuclear, wind, tidal, and solar power). In addition, carbon sequestration can be used to serve the purpose. Carbon sequestration involves the long-term storage of carbon dioxide underground, as well as on the surface of Earth in forests and oceans. Carbon sequestration in forests and oceans relies on natural processes such as forest growth. However, the clearing of forests for agricultural and other purposes (and also the pollution of oceans) diminishes natural carbon sequestration. Storing carbon dioxide underground—a technology under development that is also called geosequestration or carbon capture and storage—would involve pumping the gas directly into underground geologic "reservoir" layers. This would require the separation of carbon dioxide from power plant flue gases—a costly process.

Air Pollution Control Equipment

Air pollutants are generated by nearly every facet of the industrial process, including raw material sourcing, product manufacturing, maintenance and repair services, and distribution. Consequently, there are several different types of air pollution control equipment available for air pollutants produced by both mobile and stationary sources across a wide range of industries.

In an industrial setting, air pollution control equipment is an umbrella term referring to equipment and systems used to regulate and eliminate the emission of potentially hazardous substances—including particulate matter and gases—produced by manufacturing, process system, and research applications into the air, atmosphere, and surrounding environment. Control equipment has applications in a wide range of industries, preventing the release of chemicals, vapors, and dust and filtering and purifying the air within the work environment. Typically, fans or blowers direct industrial exhaust and emissions into the air pollution control equipment and systems which remove or reduce air pollutants through the use of one or more of the following processes:

- Combustion (i.e., destroying the pollutant).

- Conversion (i.e., chemically changing the pollutant to a less harmful compound).

- Collection (i.e., removing the pollutant from the waste air before its release into the environment and atmosphere).

Some types of air pollution control equipment applied to industrial applications and which utilize one or more of the methods of air pollutant removal or reduction mentioned above include:

- Scrubbers

- Air Filters

- Cyclones

- Electrostatic Precipitators

- Mist Collectors

- Incinerators

- Catalytic Reactors
- Biofilters

Scrubbers

Some of the most commonly used air pollution control devices in manufacturing and processing facilities, industrial air scrubbers employ a physical process—i.e., scrubbing—which removes particulates and gases from industrial emissions, such as smokestack exhaust (in the case of exhaust air scrubbers), before they are released into the atmosphere. There are two main categories of scrubbers—dry scrubbers and wet scrubbers.

Dry Scrubbers

Dry scrubbers, also referred to as dry adsorption scrubbers, inject dry, neutralizing chemical agents, such as sodium bicarbonate, into the emission stream, causing the gaseous pollutants contained within to undergo a chemical reaction which either neutralizes the pollutants or converts them into innocuous substances. Once the chemical reaction concludes, filters within the scrubber chamber collect and remove the spent agents from the cleaned emission gas. In some cases, the collected agents can be washed and reused for future dry scrubbing processes, but, if not possible, the scrubbing waste must be disposed of by specialists. Typically, dry scrubbers are used to remove or counteract acid gas within industrial emissions. The chemical reactions resulting from the addition of neutralizing agents during the dry scrubbing process helps to both reduce the acidity of the emissions and remove air pollutants.

Wet Scrubbers

Wet scrubbers, also referred to as wet adsorption scrubbers or wet collectors, employ liquid solutions—typically water—to collect and remove water-soluble gas and particulate pollutants from industrial emissions. The wet scrubbing process either passes a gas stream through a liquid solution or injects a liquid solution into a gas stream. As the gas stream comes into contact with the liquid, the solution absorbs the pollutant removing it from the stream. The types of wet scrubbing equipment available include venturi, packed bed (or packed tower), and bubbling scrubbers.

Air Filters

Air filters are devices used to control air pollution which employ a specific type of filtration media—e.g., fabric, sintered metal, ceramic, etc.—to collect and remove dry particulates and contaminants, such as dust, pollen, microbes, chemicals, etc. from air passing through them. These devices are utilized in residential, commercial, and industrial applications to remove pollutants from exhaust air and improve the air quality within the work environment. For industrial applications, there are several types of air filters available, including HEPA filters, fabric filters, and cartridge dust collectors.

HEPA Filters

Industrial HEPA filters, also known as high-efficiency particulate air filters, are a category of air filters which employ fiberglass filter mats to mechanically remove airborne particulates, such as pollen, smoke, dust, and bio-contaminants, from within the work environment. Typically the fiberglass filter mats have fibers ranging in size between 0.5 to 2 µm. When the blower component of the filtration system passes air through the HEPA filter, particulates adhere to or become embedded within the fibers. Additionally, particles passing through the filter collide with the gas contained within, which slows their velocity and increases their chance of becoming adhered to or embedded in the filter.

An example of a HEPA air filter mat.

To be classified as a HEPA filter, the United States Department of Energy (DOE) designates that a filtration system must be designed to maintain a 99.97% efficiency for collecting and removing particulates greater than or equal to 0.3 µm in diameter. Despite their high efficiency, HEPA filters are typically used in conjunction with other filtration components and systems to filter and purify the air further, such as ultraviolet light, ionizers, and activated carbon air filters.

Fabric Filters

Fabric filters—also referred to as baghouses—are a category of air filters which typically employ cylindrical fabric bags to trap and remove airborne dust and other particulates. As air passes through a baghouse, the particulates collect and accumulate on the filter's surface. The accumulation increases the efficiency of the filter, allowing smaller particles to be collected and causing a buildup of pressure across the filter fabric. Some baghouses are capable of attaining 99.9% efficiency, even for small particulate matter. These types of filters are suitable for filtering out air pollutants in a variety of industrial processes, including power plants, metal processing centers, and foundries, as well as part of multi-stage cleaning systems.

The particulate accumulation and the resulting pressure differential prompt the need for periodic cleaning. There are several methods employed by baghouses to remove the accumulation from the filter bags, including:

- Shaking the filter bags.

- Introducing an air flow into the filter bags in the opposite direction to the filtration process.

- Pulsing compressed air into the filter bags in the opposite direction to the filtration process.

For any of the above-mentioned removal methods, particulates fall from the filter fabric to the bottom of the baghouse enclosure into a collection hopper for subsequent processing and disposal.

Cartridge Dust Collectors

Similar to baghouses, cartridge dust collectors are air filters which utilize cartridge filters, rather than filter bags, to collect and remove airborne dust and other particulates. The particulate accumulation on cartridge dust collectors also requires periodic cleaning and removal. While this type of dust collector can attain higher efficiencies than a comparable baghouse, depending on the cartridge material, it can also be more sensitive to condensation.

Cyclones

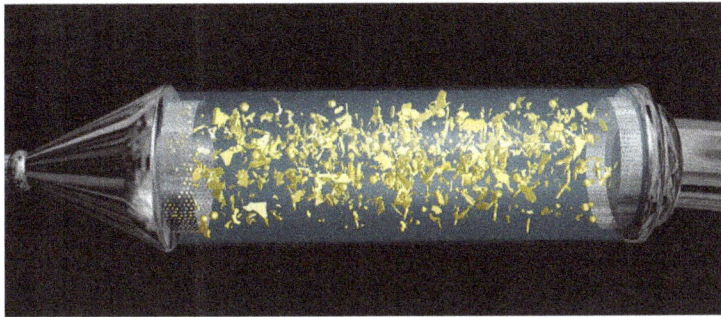

A 3D rendering of a cyclone dust collector separating particulate matter from a gas stream.

Cyclones, also referred to as cyclone dust collectors, are air pollution control devices which, similarly to air filters, separate dry particulate matter from gaseous emissions. However, rather than employing filtration media, cyclones utilize centrifugal force to collect and remove particulates. As gas streams enter a cyclone, they flow along a spiral path within the cylindrical chamber. This swirling motion forces large particulates against the chamber wall, which slows their inertia, causing them to drop into the collection hopper below for further processing and disposal. The cleaned gas streams continue upward and out of the cyclone.

While cyclones are typically employed for filtration applications of particulates ≥ 50 µm in diameter, some models are capable of greater than 90% efficiency for particulates ≥ 10–20 µm in diameter. Efficiency increases or decreases depending on larger or smaller particulate diameters, respectively. Typically, additional filtration devices used to control air pollution, such as baghouses, are employed following cyclones in an air pollution control system to remove the smaller particulates not previously separated and collected from the gas stream by the cyclones.

Electrostatic Precipitators

Electrostatic precipitators (ESPs), like air filters and cyclones, are air pollution control devices used to collect and remove particulate matter, such as dust, from industrial emissions and exhaust. ESPs employ transformers to create high static electrical potential difference between charging electrodes and collecting plates. As gas streams pass between the two components, an electrical charge is introduced to the particulates, which attracts the particulate matter to the collecting plates. Similarly to air filters, PM accumulation is periodically removed from the collecting plates and deposited in a collection hopper below, either through mechanically dislodging the particulates or by introducing water to clean off the particulates. ESPs which employ the latter method are known as wet ESPs. As ESPs typically have multiple collection plates, their efficiencies often exceed 99%.

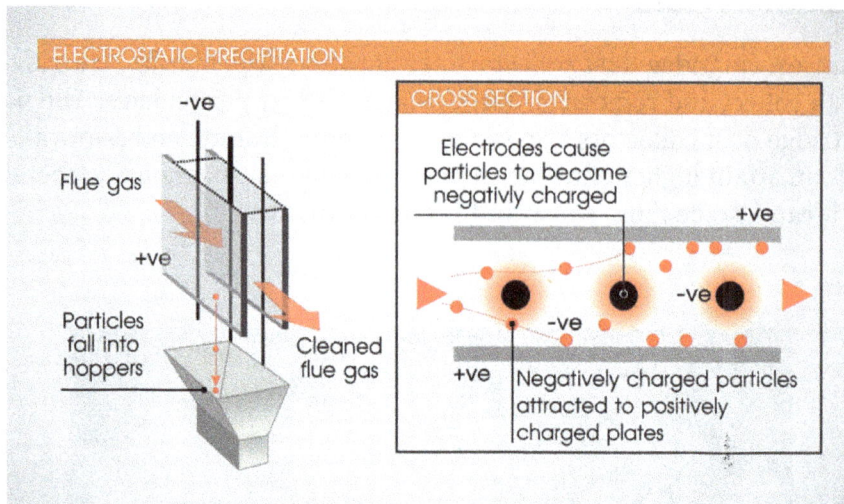

A diagram illustrating the mechanism of an electrostatic precipitator.

Mist Collectors

Mist collectors, also known as mist or moisture eliminator filters, are air pollution control devices which remove moisture and vapor—e.g., smoke, oil, mist, etc.—from gas streams. These devices employ fine mesh-like filters to separate liquid droplets from the gas and collect them into a separate chamber for further processing and, potentially, recovery and reuse.

Mist collectors maintain high filtration efficiencies for submicron liquid particles, with some models offering 99.9% efficiency for particles ≥0.3 µm in diameter. While mist collectors are capable of processing acidic and corrosive gas streams, they cannot handle gas streams containing large particulates, as they may cause an obstruction within the collector's filter. They also are not used in applications which have temperatures above 120 °F.

Incinerators

Incinerators are devices which employ combustion methods to break down pollutants into non-toxic byproducts. While these devices can be used to incinerate wastes in solid, liquid, and gaseous form, they are widely employed in a variety of industrial applications to maintain air quality and regulate gas emissions by converting VOCs, hydrocarbons, and other hazardous air pollutants (HAP) into innocuous compounds, such as carbon dioxide, nitrogen, and oxygen. Typically, incinerators are succeeded by scrubbers in an air pollution control system, as the scrubbing process removes any additional compounds formed through the combustion process.

Depending on the composition of the waste product, the incineration process can be either self-sustaining or requires supplementary fuel to ensure complete combustion of the waste compounds. Additionally, some incinerator models are available with regenerative and recuperative capabilities and are suitable for both continuous and batch applications. There are several types of incinerators available, including thermal oxidizers and catalytic oxidizers.

Thermal Oxidizers

Thermal oxidizers, also known as thermal incinerators, are incineration devices which employ a combustion process to convert particulate matter and gaseous pollutants, such as VOCs, hydrocarbon compounds, and odorous fumes, into water vapor, carbon dioxide (CO_2), and waste heat. Some types of thermal oxidizers are capable of incinerating air pollutants at 99.99% efficiency.

The types of thermal oxidizers available include direct fired thermal oxidizers (also known as afterburners), regenerative thermal oxidizers (RTOs), and recuperative thermal oxidizers. The suitability of each type of thermal oxidizer for an incineration application is dependent on the requirements of the application. For example, afterburners are best suited for applications with high concentrations of VOCs and require low capital expenditure, but do not offer options for heat recovery. On the other hand, RTOs, while more expensive than afterburners, offer capabilities for 97% heat recovery efficiency which make them more suitable for low concentrations of VOCs and continuous incineration operations.

Catalytic Oxidizers

Catalytic oxidizers are incinerators which employ catalyst beds to aid the incineration process for gaseous pollutants and particulate matter. Made of precious or base metal, the catalyst bed lowers the required temperatures for initiating oxidation, accelerating the process and reducing the amount of combustible compounds needed to achieve combustion efficiencies comparable to that of thermal oxidizers. Like thermal oxidizers, catalytic oxidizers are used in industrial applications to break down VOCs, hydrocarbon compounds, and odorous fumes. Some catalytic oxidizers are not suitable for incinerating gas and emissions containing PM as the particulates can coat the surface of the catalyst bed, preventing and disrupting the oxidation process, newer catalysts allow catalytic oxidizers to handle most gas and PM compounds.

The types of catalytic oxidizers available include regenerative catalytic oxidizers (RCO) and recuperative catalytic oxidizers.

Catalytic Reactors

Catalytic reactors, also referred to as selective catalytic reduction (SCR) systems, are air pollution control devices widely used to mitigate nitrogen oxide (NO_x) emissions produced by the burning of fossil fuels in industrial applications. These devices first inject ammonia into the industrial exhaust and emissions, which reacts with the NO_x compounds to produce nitrogen and oxygen. Similarly to incinerators, these devices also employ other catalysts which enable some of the remaining gaseous pollutants to undergo combustion for further processing and reduction. One common application of catalytic reactors is in modern automobiles; the three-way catalytic converter in a car's exhaust system is used to reduce the amounts of NO_x, CO, and other VOCs in the engine emissions.

While for NO_x reduction and removal, SCR systems can achieve more than 90% efficiencies, for other gaseous pollutants these devices can achieve 99.99% efficiencies with lower energy requirements compared to incinerators. Despite the high efficiencies possible, SCR systems are not suitable for all gaseous pollutant reduction applications as the large amounts of catalyst required are costly, and the systems cannot process emissions and exhaust containing particulate matter.

A close-up of a catalytic converter in an automobile.

Biofilters

Biofilters are air pollution control devices which employ microorganisms, such as bacteria and fungi, to degrade and remove water-soluble compounds. Similarly to incineration devices, biofilters destroy the pollutants to reduce the amount present in industrial emissions and exhaust. However, the microorganisms in biofilters absorb and metabolize gaseous pollutants, such as VOCs and organic HAP, without generating byproducts typically produced through combustion, such as NO_x and CO. These devices are capable of achieving over 98% efficiencies.

Auxiliary Equipment

A complete air pollution control system refers not only to the equipment which destroys, chemically changes, or collects the pollutants and emissions—e.g., scrubbers, filters, ESPs, incinerators, etc.—but the components and equipment which make up the system's infrastructure and provide support to the control equipment as well. Some of the standard auxiliary components and equipment includes:

- Monitoring and Pollution Control Systems: Monitoring and control systems, such as continuous emission monitoring systems (CEMS), allow companies to monitor, control, and record their emission levels. By doing so, they can better track their emissions output and system efficiency, as well as update their systems accordingly to better align with environmental or budgetary standards.

- Fans and Blowers: Fans and blowers are incorporated into air pollution control systems to help draw and direct industrial exhaust and emissions into the air pollution control equipment for pollutant filtration and removal, as well as to help guide the clean, filtered air out of the equipment.

- Blade dampers: Blade dampers can be used to control and regulate air flow within air pollution control systems. These devices are available in manually or automatically-operated designs, as well as employ electrical or pneumatic actuators to open and close the device's blades. There are two main types of dampers available—parallel and opposed blade dampers—each of which is suited for different specifications and requirements. Table below illustrates the characteristics of each type, including the advantages, disadvantages, and suitable applications.

Table: Characteristics of Dampers by Type.

Type of Damper	Characteristics
Parallel Blade Dampers	• Blades open in the same parallel direction • Rapid increase in flow rate as blades open • Sensitive to arm swings • Used for open/close applications • Simpler design, more cost-effective • Limited suitability: only between 75%–100% open
Opposed Blade Dampers	• Blades open in opposite directions • Gradual increase in flow rate as blades open • Used for modulating applications • Broader suitability: between 25%–100% open • More complex design, less cost-effective

- Stacks: Once industrial emissions are processed by the air pollution control equipment employed, the processed emissions are generally released into the atmosphere. Stacks provide an outlet through which the emissions are dispersed into the atmosphere, as well as prevent the emissions from re-entering the building. Additionally, stacks produce a natural draft which aids, when applicable, combustion processes.

- Heat Exchangers: Heat exchangers can be used within waste heat recovery systems. By employing waste heat recovery systems, air pollution control systems can improve the energy and operating efficiency of their equipment. This characteristic is especially useful for control devices which employ the combustion method of pollutant removal, such as thermal oxidizers.

Pollution Control Systems

While there are a wide variety of air pollution control equipment and systems available, the suitability of each type in mitigating the amount of chemicals, vapors, and dust emitted into the atmosphere and filtering and purifying the air within the work environment is dependent on several factors. Some of the considerations that industry professionals should keep in mind when choosing an air pollution control device include:

- The type(s) of air pollutant(s) needing removal.

- Pollutant removal and reduction efficiencies.

In regards to air pollution control equipment, efficiency refers to the amount of emissions collected, controlled, reduced, or eliminated by a device represented by a percentage value. Typically, when comparing air pollution control devices, industry professionals consider the capture efficiency—i.e., the percentage of emissions gathered and directed to a control device—and the control efficiency—i.e., the percentage of air pollutants removed from an emission stream under ideal conditions. The control efficiency value can be represented by the following equation:

$$(1-\frac{\text{Uncontrolled Pollutant Emissions Rate- ControlledPollutant Emissions Rate}}{\text{Uncontrolled Pollutant Emissions Rate}})*100$$

Whereas the uncontrolled pollutant emission rate represents the total value amount of the pollutant concentration within emissions and exhaust produced by an industrial application multiplied by the volumetric flow rate, the controlled pollutant emission rate represents the amount of pollutant removed from the emissions and exhaust by the air pollution control device.

While control efficiencies provide a metric of the maximum amount of pollutants which can be controlled by a device, a variety of conditions and circumstances can affect a device's overall (i.e., actual) efficiency, including the capture efficiency, age and condition of the device, type(s) of pollutant(s) and its/their properties, pollutant stream flow rates and concentrations, temperature, humidity, and device surface area and volume. While changes to some factors can lead to decreases in overall efficiency, changes to others can lead to increases in efficiency. For example, overall efficiency can be represented by the following equation:

$$\text{Overall control Efficiency[\%]} = \text{Capture Efficiency[\%]}*\text{control Efficiency[\%]}$$

If the capture efficiency value worsens, overall control efficiency decreases, whereas if the capture efficiency improves, overall control efficiency increases.

Following are some types of control equipment and their device considerations and restrictions:

Scrubbers

Dry Scrubbers

- Gas streams should be cooled and diluted first to allow for optimal pollutant removal conditions.

- Spent reagents require specialized disposal.

Wet Scrubbers

- Liquid reagent must come into direct contact with pollutant to allow for removal.

- Optimal type determined by pollutant being removed (gas or PM); e.g., PM is removed with spray towers scrubbers, and gas is removed with packed bed scrubbers.

- For PM: smaller particle size results in lower efficiency.

Air Filters

HEPA Filters

- High temperatures or humidity can damage filtration media.

- Requires periodic changes of filtration media, which cannot be cleaned.

- Used filters can result in secondary waste accumulation.

Fabric Filters (Baghouses)

- Abrasion, high temperatures (>290 °C), and chemicals can damage filtration media.

- Requires periodic cleaning of filtration media.

- Available with catalytic filters bags for additional chemical filtration.

- High resistivity PM does not affect efficiency.

Cartridge Dust Collectors

- Filtration media can be affected by high humidity.

- Requires periodic cleaning of filtration media.

Cyclones

- Smaller particle size results in lower efficiency.

- Capable of handling high gas stream temperatures.

- Generally used in conjunction with other air pollution control devices.

Electrostatic Precipitators

- Cannot handle high or low resistivity particles (requires moderate resistivity).

- PM size does not significantly affect efficiency.

Mist Collectors

- Cannot handle gas streams with large particulate matter.

- Limited to handling gas stream temperatures <120 °F.

Incinerators

Thermal Oxidizers

- Not ideal for halogen or sulfur compounds due to the formation of corrosive gases.

Catalytic Oxidizers

- Older catalysts cannot handle gas streams with PM.

Catalytic Reactors

- Catalyst contamination can reduce efficiency.

- Cannot handle gas streams with PM.

Biofilters

- Pollutants must be water-soluble.

In addition to the type of air pollutant needing removal and the control efficiency of an air pollution control device, industry professionals may also consider other device characteristics when choosing equipment for their air pollution control application.

Devices used to Control Air Pollution

Smoke and exhaust emitted from industrial chimneys.

Air pollutants, both gaseous compounds and particulate matter, are generated by nearly every facet of the industrial process across a wide range of industries. Therefore, air pollution control equipment is available for almost every of industrial applications, including agriculture, automotive, pharmaceuticals, metallurgy, wastewater treatment, oil and gas, and power production.

Following are listed some types of control equipment and their common industrial applications:

Scrubbers

Dry Scrubbers

- Combustion processes,

- Soil remediation,

- Oil refineries,

- Wastewater treatment plants,

- Paint, powder coating, and finishing shops,

- Metallurgical plants.

Wet Scrubbers

- Combustion processes,

- Power plants,

- Construction material manufacturing,

- Fertilizer plants,

- Chemical manufacturing and processing.

Air Filters

HEPA Filters

- Cleanroom and research facilities,

- Hospitals,

- Pharmaceuticals,

- Computer and electronics manufacturing,

- Aerospace,

- Nuclear power plants.

Fabric Filters (Baghouses)

- Combustion processes,

- Power plants,

- Metallurgical plants,

- Foundries,

- Fertilizer plants,

- Pharmaceuticals,

- Automotive,

- Mining,

- Construction material manufacturing.

Cartridge Dust Collectors

- Combustion processes,
- Automotive,
- Bulk processing,
- Chemical and pharmaceutical manufacturing,
- Machining, fabrication, and finishing shops,

Cyclones

- Cotton gins,
- Aggregate and construction material manufacturing,
- Metallurgical plants,
- Woodworking shops,
- Pharmaceuticals,
- Food and chemical processing,
- Combustion processes.

Electrostatic Precipitators

Mist Collectors

- Metal machining and finishing shops,
- Food and chemical processing,
- Desalination plants,
- Paper and pulp mills,
- Agriculture,
- Marine.

Incinerators

Thermal Oxidizers

- Landfills,
- Oil and gas refineries,

- Chemical manufacturing,

- Rubber and plastic manufacturing,

- Coating and printing shops.

Catalytic Oxidizers

Catalytic Reactors

- Automobiles,

- Combustion processes,

- Landfills,

- Oil refineries,

- Coating and printing shops.

Biofilters

- Wastewater treatment,

- Food processing,

- Agriculture,

- Oil and gas refineries,

- Pharmaceuticals.

Low-cost Air Pollution Sensors

Low-cost air pollution sensors are attracting more and more attention. They offer air pollution monitoring at a lower cost than conventional methods, in theory making air pollution monitoring possible in many more locations.

Measurements with low-cost sensors are often of lower and more questionable data quality than the results from official monitoring stations carried out by EU Member States in accordance with European legislation and international standards methods.

If the quality of the measurements can be improved, sensors could become a game changer in monitoring air pollution, traffic management, personal exposure and health assessment, citizen science and air pollution assessment in developing countries.

Electrochemical sensors are based on a chemical reaction between gases in the air and the electrode in a liquid inside a sensor.

In a metal oxide sensor (resistive sensor, semiconductor) gases in the air react on the sensor surface and modify its resistance.

A photo ionization detector ionises volatile organic compounds and measures the resulting electrical current.

Optical particle counters detect particulate pollution by measuring the light scattered by particles.

Optical sensors detect gases like carbon monoxide and carbon dioxide by measuring the absorption of infrared light.

Metal Oxide Sensors (Used to Measure NO_2, O_3, CO)

- Low cost: around 10 - 15 € for a sensor.

- Good sensitivity, from mg/m^3 to $\mu g/m^3$.

- Results are affected by temperature and humidity variations.

- Long response time (5 – 50 min).

- Output depends as well on history of past inputs.

- Instability can be observed.

Electrochemical Sensors (Used to Measure NO_2, SO_2, O_3, NO, CO)

- Medium cost: around 50 - 150 € for a sensor.

- Good sensitivity, from mg/m³ to µg/m³.

- Fast response time (30-200s).

- Highly sensitive to temperature and humidity variations (change in meteorology) depending on electrolyte.

- Selectivity: show cross-reactivity with similar molecule types.

Photo Ionization Detector (Used to Measure VOC)

- Moderate price: 400 € for a sensor to 5000 € for handled device.

- Good sensitivity, down to mg/m3, some down to µg/m³.

- Limited temperature dependence and humidity effects.

- Very fast response time (few seconds).

- Not selective: reacts to all VOCs that can be ionised by the UV lamp.

- Significant signal drift.

Optical Particulate Counter (Used to Measure PM)

- Moderate cost: 300 € for a sensor to 2000 € for handled device.

- Fast response time (1 s).

- Sensitivity in the range of 1 µg/m³.

- Able to identify the size of the particle (PM10, PM2.5,..).

- Conversion from particle counts to PM mass is based on theoretical model.

- The measured signal depends on a variety of parameters such as particle shape, color and density, humidity, refractive index.

Optical Sensors (Used to Measure CO, CO_2)

- Moderate cost: 100 – 350 € for sensor to 2000 € for handheld device.

- Good sensitivity for CO_2 (350 – 2000 ppm).

- Selectivity is good through characteristic CO_2 IR spectra.

- Response time 20 – 120 s.

- Limited drift over time of the sensor calibration.

- Need for correction for the effects of temperature, humidity and pressure.

What Data Quality to Expect from Sensors?

The signals from sensors not only depend on the air pollutant of interest, but also on a combination of several effects, such as other interfering compounds, temperature, humidity, pressure and signal drift (instability of signal). At high concentrations the signal from the air pollutant can be strong, but at ambient air levels the signal is weaker in comparison to the interfering effects.

The quality of sensor results therefore depends on technology and implementation (application, site, conditions, set-up). Reproducing sensor responses at different measurement sites or the portable use of sensors is thus difficult. Due to the influence of meteorological parameters on a sensor signal, simple correction and/or calibration is not always possible.

Nevertheless, in certain well-defined situations, the measurement uncertainty of these devices may approach the level of 'official' measurement methods.

Why do Measurements from a Sensor and from a Monitoring Station Differ?

An air pollution analyser inside an official monitoring station uses a well-defined, standardised and selective principle. Analysers are type approved and tested for interferences and under varying conditions. The environment in official monitoring stations is controlled, their instruments are regularly checked, and the measurements are subject to rigorous quality control and calibration procedures.

Some sensors can be sensitive to weather conditions (wind speed, temperature, humidity) or can have difficulties distinguishing pollutants. When using sensors, the measurements should be carefully evaluated and validated.

The spatial representativeness of a measured pollutant concentrations depends on pollutant, source and surroundings. Even if a measurement is carried out correctly, it may only be representative for a very small area.

A protocol to evaluate sensors based on common criteria is under development by the European Standardisation Organisation (CEN).

Ways to Stop Air Pollution

- Conserve energy - at home, at work, everywhere.

- Look for the ENERGY STAR label when buying home or office equipment.

- Carpool, use public transportation, bike, or walk whenever possible.

- Follow gasoline refueling instructions for efficient vapor recovery, being careful not to spill fuel and always tightening your gas cap securely.

- Consider purchasing portable gasoline containers labeled "spill-proof," where available.

- Keep car, boat, and other engines properly tuned.

- Be sure your tires are properly inflated.

- Use environmentally safe paints and cleaning products whenever possible.

- Mulch or compost leaves and yard waste.

- Consider using gas logs instead of wood.

On Days when High Ozone Levels are Expected, Take these Extra Steps to Reduce Pollution:

- Choose a cleaner commute - share a ride to work or use public transportation.

- Combine errands and reduce trips. Walk to errands when possible.

- Avoid excessive idling of your automobile.

- Refuel your car in the evening when it's cooler.

- Conserve electricity and set air conditioners no lower than 78 degrees.

- Defer lawn and gardening chores that use gasoline-powered equipment, or wait until evening.

On Days when High Particle Levels are Expected, Take these Extra Steps to Reduce Pollution:

- Reduce the number of trips you take in your car.

- Reduce or eliminate fireplace and wood stove use.

- Avoid burning leaves, trash, and other materials.

- Avoid using gas-powered lawn and garden equipment.

References

- Air-pollution-control, technology: britannica.com, Retrieved 19 May, 2019

- Understanding-air-pollution-control-equipment, plant-facility-equipment: thomasnet.com, Retrieved 21 July, 2019

- Brochure%20lower-cost%20sensors, air, environment: europa.eu, Retrieved 28 April, 2019

- Reducepollution, airquality, region1: epa.gov, Retrieved 1 August , 2019

Permissions

Index

www.ingramcontent.com/pod-product-compliance
Lightning Source LLC
Chambersburg PA
CBHW082013190326
41458CB00010B/3179